THE MICROBES,
OUR
UNSEEN FRIENDS

THE MICROBES, OUR UNSEEN FRIENDS

BY HAROLD W. ROSSMOORE

Wayne State University

Wayne State University Press
Detroit, 1976

Second printing, November 1977.

Library of Congress Cataloging in Publication Data

Rossmoore, Harold William, 1925–
 The microbes, our unseen friends.

 Bibliography: p.
 Includes index.
 1. Microbiology. 2. Food—Microbiology.
3. Industrial microbiology. I. Title.
QR56.R595   576'.16   76-17795
ISBN 0-8143-1561-5

Waynebook 42

This book is dedicated, with love, to my wife, Shirl, who taught me how to listen and who persevered when I didn't.

And to all my kids who helped me grow up and kept me from growing old.

ACKNOWLEDGMENT

This book was helped along considerably by many people. My first editor, Dick Boolootian, I must thank for encouragement in the selection of the subject matter. I owe a debt of gratitude to Bill Mayer, who approved the subject matter for this book in spite of his close acquaintanceship with the author. To Ed Kormondy I am also indebted, for without his work on the manuscript the final writing would have been much more painstaking.

I would also like to thank Misses Sadusk, Cardew, Switzer, Ferguson, and Melendez for their yeoman contribution to the translation of my writing to finished type.

And lastly, I would like to thank all my unseen friends, wherever they are.

CONTENTS

ILLUSTRATIONS

TABLES

Throughout the book measurements are given in the English or the metric system, as appropriate. Equivalents are as follows:

$$\frac{5 \times (\text{Degrees Fahrenheit (F°)} - 32)}{9} = \text{Degrees Celsius (C°)}$$

1 gallon = 3.8 liters (3800 milliliters)
(128 liquid ounces)

1 ounce
(avoirdupois) = 30 grams

PROLOGUE:
ANIMALCULES, MICROBES,
AND
FILTRABLE VIRUSES

This is a prologue because its intent and content is different
from the rest of the book. Indeed, one definition of a prologue
is, "the opening scene of a play whose main action is set within
a separate frame." In the following eight chapters I bring to you
as many of the useful activities of microbes as could be conveniently contained in a book of this size. Quite often I have had
to be technical. After six chapters had been completed, my
wife, my editor, Ed Kormondy, and even I realized that without
a "cast of characters" many of my readers would not know
who was playing what role. So in a sense this prologue is a
recapitulation of the characters who appear later on, presented
as briefly as possible. The following is not intended as a short
course in microbiology; it replaces a glossary, to which I have a
strong personal aversion. If you choose, you can skip this pre-
amble and proceed to Chapter I. In the event that you are
sufficiently perplexed by the names and the terms you come
across, you can always turn back to the prologue.

WHAT ARE MICROBES

In defining *microbes,* we owe a debt to Anthony Van Leeuwenhoek, a Dutch dry goods merchant who was also an amateur naturalist and lens grinder. Beginning in 1674, he began a series of reports to the Royal Society of London on the strange tiny creatures he was observing in teeth scrapings, rain water, and feces, among other things. Leeuwenhoek wrote, "for I noticed one of my back teeth, up against the gum, was coated with the said matter for about the width of a horse-hair, where to all appearance, it had not been scoured by the salt for a few days; animalcules here, that I imagined I could see a good 1,000 of 'em in a quantity of this material that was no bigger than a hundredth part of a sand-grain."[1] His sketches and his comparisons in size with blood corpuscles have convinced latter-day microbiologists that he was the first one ever to see organisms called bacteria (Fig. 1). Leeuwenhoek was such an especially gifted craftsman that neither his contemporaries nor succeeding generations of lens grinders could duplicate his lenses (Fig. 2, L. microscope). Thus, it was not until two hundred years later that his "animalcules" were again seen and studied in earnest.

Although a variety of different kinds of microorganisms had been described by the middle of the nineteenth century, their orderly segregation into systematic groups was limited due to the inability to examine them in any real detail. At that time true *Yeasts* were well known, as well as one-celled animals called *Protozoa,* and what we refer to as *molds.* With the refutation of the theory of spontaneous generation and the demonstration of causes of fermentation by Pasteur, as well as Robert Koch's elegant proof of the germ theory of disease, more and more, individuality became distinguishable among microbes.

In 1878 Pasteur gave his approval and acceptance to the word *microbe,*[2] despite the fact that etymologically it means "short-lived thing," not "very little thing". Because of and in spite of their exactness, scientists are no different than Humpty-Dumpty.[3]

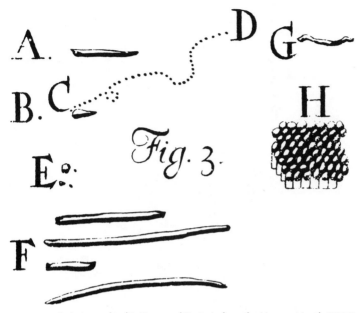

1. Leeuwenhoek's Figures of Bacteria from the Human Mouth (1684)

WHAT'S IN A NAME!

As more and more diseases were attributed directly to microbes, each disease had associated with it a specific organism. Because of this and other related research activities, for example in soil and spoilage, the science of microbiology developed a very sound practical base; techniques and methodology kept pace with the discovery of new microorganisms. Bacteria, a kind of microbe (from the Greek, meaning "little stick"), were in number the more exciting new discovery.

In 1892, a Russian biologist, Dimitri Ivanowsky, deduced that a disease of tobacco called tobacco mosaic was transmissible by an active principle; that is, it seemed contagious. A contemporary, Martinus Beijerinck (the founder of the Delft School of Microbiology in the Netherlands), confirmed in 1898, that the cause of the disease was living and transmis-

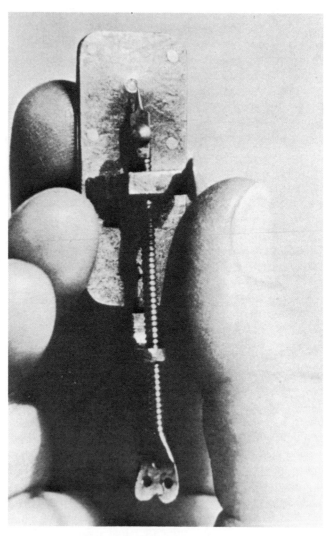

2. Leeuwenhoek's "Microscope"
Replica courtesy of Archives of the American Society for Microbiology.
Reproduced by permission of the American Society for Microbiology.

sible. He ground up sick tobacco leaves, separated the juice and filtered it through unglazed porcelain with pores that prevented the passage of any organisms that could be seen with available microscopes (about 1/100,000 inch). This filtrate, even diluted, when applied to healthy tobacco leaves not only caused disease but did so in the original intensity, suggesting that something was multiplying. Beijerinck first termed the principle *virus*, Latin for poison, and synonymous with toxin or infectious agent. Since it passed through a filter that held back all known and visible forms, it was called "*filtrable* virus."[4] Subsequently, the "filtrable" was dropped from its name, but filtration in some form was the major means of detecting and sizing viruses until 1939, when they were first seen with the electron microscope. (The tobacco mosaic virus was the first to be seen.)

TAXONOMY—A NECESSARY EVIL
Taxonomy is the science of classification: "a place for everything and everything in its place." Man's penchant for pigeonholing is often thwarted by Mother Nature, whose creatures sometimes fit no pigeonholes. This is particularly true with the catchall *microbe*.

By strictly referring to size in calling an organism *microbe,* we cut across many familial boundaries. In other words, some microscopic forms are called animals, some are called plants, and others cause confusion. While one needs no advanced biological training to ascertain that a tree is a plant or a dog is an animal, the less specialized the zoological and botanical representatives, the more difficult it becomes to make a choice. Three criteria have been used to help make that choice. They are: the cell wall, independent movement, and type of nutrition (Table 1). You can readily see that what we call microorganisms are not easily placed in one or the other camp. Nevertheless, by convention the presence of a cell wall is used to separate the two groups. Thus, protozoa are recognizable as

TABLE 1
Whether Plant or Animal

	Cell Wall	Independent Movement	Type of* Nutrition
Plant	yes	no	autotroph
Animal	no	yes	heterotroph
Protozoa	no	yes	heterotroph
Algae	yes	yes & no	autotroph
Molds	yes	no	heterotroph
Yeasts	yes	no	heterotroph
Bacteria	yes (95%)	yes & no	autotroph & heterotroph
Viruses	no	no	heterotroph

*Autotroph literally means "self-nourishing," implying that organisms so-called make their own food. All green plants derive their energy from sunlight and appear to make their cell constitutents with no other help. Heterotroph means "nourished by others," and organisms that in some way digest or absorb nutrients manufactured by other living things are so classified. Although these terms are convenient ones, they are not completely correct since, for one, green plants cannot grow with solar energy alone. They must be given "plant food" (i.e., fertilizer). Self-nourishing, then, means only "production of sugar from carbon dioxide."

animals, and yeasts and molds (fungi) as plants, as are bacteria (but with greater reluctance). The viruses are a special case, which I will discuss later. (I make no brief for the above separation, or for one that establishes a third kingdom, the Protista, which excepts all one-celled and undifferentiated multicelled organisms from the plant and animal kingdoms. However, a beginning must be made.)

FUNGI, YEASTS AND MOLDS
The terms fungus and mold are synonymous, as is the reference to mildew. They describe (with the exception of mushrooms) a cottony or powdery growth, many with particular colors or odors easily recognized by the layman. The moldy look results from the growth of aerial filaments called hyphae (singular hypha) which en masse are referred to as mycelium. The organ-

isms so described are all plants in the phylum Eumycophyta (true fungi), and have no stems, roots, leaves, or chlorophyll. Collectively, they cause diseases of plants and animals, deterioration and spoilage, as well as the more pleasant activities described later on.

Of the four[5] major classes of fungi, three have the distinction of being able to reproduce sexually and asexually. These are the Phycomycetes (algae-like fungi), the Ascomycetes (sac-bearing fungi) and the Basidiomycetes (fungi with spores arising from a small base). The fourth class demonstrates the anthropocentric smugness of the scientist; these are called the Imperfect Fungi because they lack a perfect stage, or sex life.

The stages that we most often encounter vary with the group; for example, the asexual stages of the Phycomycetes and Ascomycetes are seen more often while the reverse is true of the Basidiomycetes (mushrooms, etc.). Since the latter class is already familiar to you, I will not discuss it further. Each of the former species has a characteristic method of asexual reproduction. In the Phycomycetes, a hypha develops into a specialized structure bearing a *sporangium* containing many spores. When the *sporangium* breaks open, the spores are released each to give rise to a new individual (Fig. 3). Perhaps the most well known of the Phycomycetes is *Rhizopus nigricans,* the black bread mold. In the Ascomycetes and Imperfects, the reproductive structures called *conidia* (literally, "dustlike") arise chainlike from the tips of specialized hyphae. *Penicillium* and *Aspergillus* genera are typical representatives.

Other asexual methods of reproduction of interest to us are found among the yeasts; these are single-celled organisms which reproduce by budding. The bud appears in synchrony with the doubling of the hereditary material in the cell (Fig. 4), half of which moves to the bud before it pinches off. Taxonomically, the yeasts are in two classes; the true yeasts are Ascomycetes. They include *Saccharomyces cerevisiae* (brewer's yeast) and the "false" yeasts which, although they have no sex

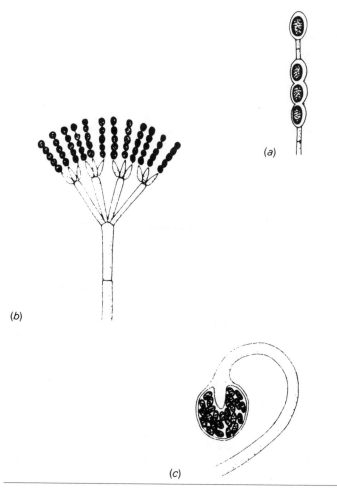

3. Different Types of Asexual Spores Formed by Fungi
(a) chlamydospores
(b) conidiospores
(c) sporangio spores
Reprinted, by permission, from Thomas Brock, *The Biology of Micro-organisms,* 2d ed. (Englewood Cliffs, N.J.: Prentice-Hall, Inc., 1974), p. 774.

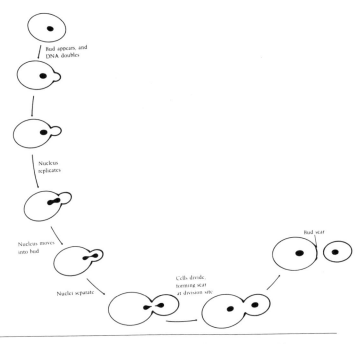

4. Cell Division Cycle in a Yeast, *Saccharomyces cerevisiae*
Reprinted, by permission, from Thomas Brock, *The Biology of Microorganisms*, 2d ed. (Englewood Cliffs, N.J.: Prentice-Hall, Inc., 1974),
p. 775.

stage, do reproduce by budding. Their colonies have the typical
shiny appearance of true yeasts.

ALGAE
Algae are undifferentiated plants without stems, roots, or leaves.
They differ from the fungi in that they do have chlorophyll.
There are almost a dozen algal phyla with different so-called
accessory pigments; thus we have green, red, brown, and blue-
green algae. Structure also differs, whether unicellular or primarily filamentous (Fig. 5). In many classification schemes, the
blue-green algae are placed in a group called the Prokaryotes
(primitive nucleus) along with the bacteria. There are a number
of valid reasons for this, the most obvious being the lack of a

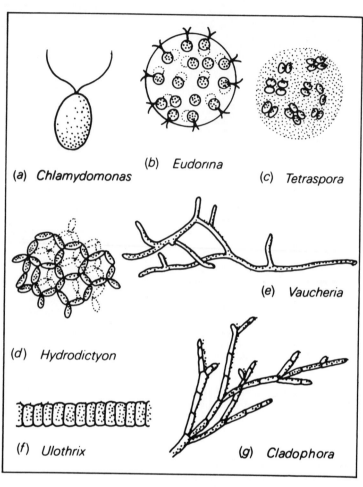

(a) Chlamydomonas
(b) Eudorina
(c) Tetraspora
(e) Vaucheria
(d) Hydrodictyon
(f) Ulothrix
(g) Cladophora

5. Various Kinds of Cell Shapes and Arrangements in Eucaryotic Algae
(a) unicellular, (b) colonial, (c) tetrasporal, (d) coenobium, (e) non-septate or siphonaceous, (f) septate (unbranched filament), (g) septate (branched filament)
Reprinted, by permission, from Thomas Brock, *The Biology of Microorganisms,* 2d ed. (Englewood Cliffs, N.J.: Prentice-Hall, Inc., 1974), p. 785.

structured nucleus with its own limiting membrane. Our main concern with algae in this book relates to their photosynthetic ability; that is, their ability to get energy from sunlight. Algae are called primary producers since they are not dependent on any other living things for energy.

Since 80 percent of the earth's surface is water, aquatic production becomes exceedingly important. For example, not only do algae produce the food that is consumed by fish, but while growing they also are prime operators in the carbon and oxygen cycles, releasing oxygen as a part of photosynthesis and taking up carbon dioxide at the same time. Many species of algae have achieved relative prominence, especially for the microbiologist. To Frau Hesse, the wife of a coworker of the famed Robert Koch, we owe the introduction of agar-agar, a polysaccharide jelly extracted from a red algae that has made more possible the isolation and study of single species of microbes on solid media. On the other hand, algae (through no fault of their own) have seen one of their virtues turned into a vice. Due to overfertilization (eutrophication) of inland waterways, some species of algae have grown luxuriously, producing so-called algal blooms. This excessive growth has succeeded in crowding out other aquatic forms and is held responsible for the pollution and potential death of our lakes. I will not dwell on this unfortunate aspect of microbial activity, but suggest that you refrain from passing final judgment until reading Chapter 7.

BACTERIA

The typical bacterial cell has a rigid, limiting wall composed of a chitinlike substance (not cellulose like higher plants), and it divides by simple binary fission (splitting into two equal daughter cells).[6] As I mentioned earlier, the presence of the cell wall has been the determining factor in relegating the bacteria to the plant kingdom. Some 3,000 species have been named and described, but fewer than 5 percent are actually involved in disease or destruction.[7] The overwhelming majority are harmless and, in fact, are beneficial (as I hope you will agree after reading the rest of the book).

Perhaps the feature that may be the most useful in characterizing an organism is size. Although there are exceptions, individual bacterial cells cannot be seen without high-powered magnification. Usually, this means using a microscope which magnifies an object 500 to 1,000 times. In fact, if an organism is too small to be seen with this kind of microscope, it probably is not a member of the bacteria (viruses cannot be seen routinely without a much more powerful device—the electron microscope. See Fig. 6.)

Bacteria are classified not only by size but also by appearance under the microscope: that is, their simple external structure. Sad to say, a large number of undergraduates passing though introductory microbiology persist in seeing eyelashes as bacteria despite the heroic efforts of their laboratory instructors. Some cannot get the knack in one academic term of looking through a high-powered microscope with one eye without being distracted by what the other eye sees. The three recognized basic bacterial shapes are (Fig. 7): spherical, cylindrical, and spiral. However, although some of the spheroids are indeed round, others are lemon- or kidney-shaped. The collective term for all the spherical bacteria is *coccus* (seed or berry). The cylindrical or rod-shaped cells (synonym *bacillus*) can be curved, boomerang fashion, as well as being straight. The spirals belong to two distinctly different groups of bacteria. In the first, cells are wavy rods with the rigid cell walls typical of bacteria. They have tufts of flagella (whiplike tails that are used for independent movement by many microbes) on one or both ends. These propel the spirals through water corkscrew fashion. In this group are found a number of photosynthetic bacteria. The other spirals are structurally much different. They belong to a group called the *spirochetes*. These have flexible walls, have no visible flagella, and move by apparently compressing and expanding body segments as in the tightening and loosening of

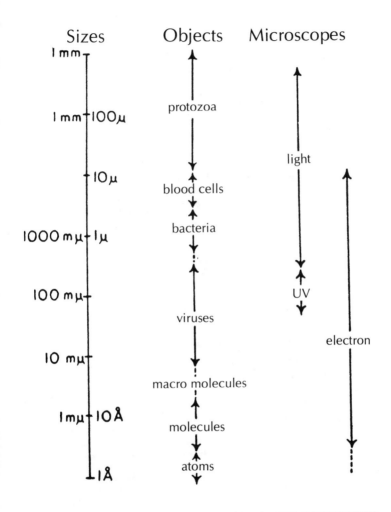

Sizes	Objects	Microscopes

6. Relative Size of Microbes, Molecules, and Atoms
Reprinted, by permission, from A.J. Rhodes and C.E. van Rooyen,
Textbook of Virology (Baltimore: The Williams & Wilkins Co., 1958),
p.34.

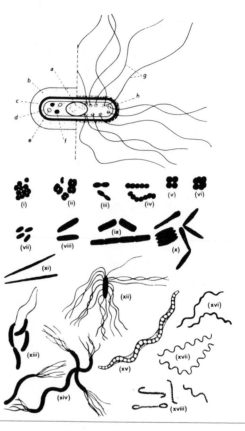

7. Anatomy and Morphology of Bacteria

A diagram showing typical anatomical structures common to many bacteria. A capsulated, non-flagellated bacillus is illustrated to the left of the dotted line and a flagellate but non-capsulate bacillus to the right.

(a) endospore (oval and central)
(b) capsule
(c) cell wall
(d) cytoplasmic membrane
(e) oil droplet or sulphur granule
(f) volutin granule
(g) flagella, each arising from a structure within the cytoplasm
(h) slime layer
(i–xviii) a variety of bacterial shapes

Reprinted, by permission, from Lilian E. Hawker, et al., *An Introduction to the Biology of Microorganisms* (New York: St. Martin's Press, 1960), chap. 1, pp. 8–9.

a spring. The most infamous member of this group is *Treponema pallidum* (the pale thread), the cause of syphilis.

In addition to the three basic forms, a number of major divisions have distinctive characteristics that set them apart. Of these, only two divisions are appropriate to future discussions in this book. One, the Actinomycetes, are very similar in appearance to the Imperfect Fungi. We are interested in them because they are the taxonomic home of many antibiotic producers, and the cause of tuberculosis and leprosy. The other, the Myxobacteria or slime-mold bacteria, are very similar in appearance to the true slime molds. These bacteria are the major responsible agents for the digestion of walls of dead cells in the soil, thus contributing to the recycling of carbon.

VIRUSES

Upon completing my discussion of viruses with my classes in introductory microbiology, I find this question invariably arises: "Are they living?" This is not an easy question to answer. Perhaps the following statement by N. W. Pirie best states the case for the virus:

> Now, however, systems are being discovered and studied which are neither obviously living nor obviously dead and it is necessary to define these words or else give up using them and coin others. When one is asked whether a filter-passing virus is living or dead the only sensible answer is, "I don't know; we know a number of things it will do and a number of things it won't and if some commission will define the word 'living' I will try to see how the virus fits into the definition." This answer does not as a rule satisfy the questioner who generally has strong but unfortunate opinions about what he means by the words living and dead.[8]

If we avoid the word *living,* how can we scientifically and adequately define the virus? I offer you a composite description from André Lwoff (Nobel laureate, 1968) and Salvador

Luria (Nobel laureate, 1970), two of the world's most eminent virologists. Viruses are strictly intracellular and potentially pathogenic entities with an infectious phase and (1) possess only one type of nucleic acid, either DNA or RNA, which are their hereditary substance; (2) are capable of infecting a specific host cell; (3) lack the enzymes required for producing useful energy; (4) do not undergo binary fission; (5) do convert the synthetic machinery of the host cell to their own ends, specifically, to make virus particles, the *virions* which contain viral genes and which are subsequently transferred to new host cells.

Despite their small size (25—200 nanometers), the electron microscope has enabled us to find out that even viruses have their own substructures, an inner core of nucleic acid and an outside wrapping of protein. This commonality all viruses have. There are other differences in shape, symmetry, and in the hosts they attack. By far the most ironic example of forced togetherness is the bacterium and the bacteriophage.[9] The latter virus was discovered in 1915 and since then has become the ultimate tool of the molecular biologist in the study of cancer and heredity. It has been called the most sophisticated of particles, for no other group of viruses has such structural complexity (Fig. 8).

HOW MICROBES GROW

Obviously, we have to know something about the particular nutritional needs of each species as well as their temperature and oxygen requirements in order to grow them. These optimal (best) conditions usually mimic those found in nature. In fact, many of the processes discussed later on like wine- and cheese-making and bread-baking reflect physiological activities of microbes naturally associated with the starting materials in those processes. The discoveries of the processes were undoubtedly prehistoric accidents, and perpetuation of these accidents means maintaining "natural conditions." This is the meaning of *ambient,* a term that is used frequently later on; for example,

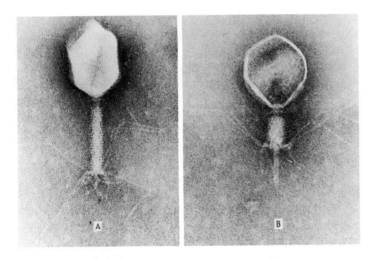

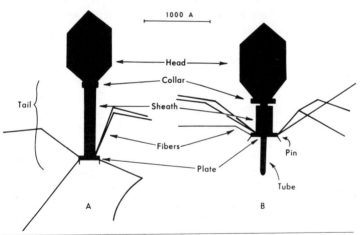

8. Structural Complexity of the Bacteriophage
Above: Micrographs of bacteriophage T2 with negative staining. Below: **A,** extended sheath; **B,** contracted sheath. In **A,** the head contains the DNA; in **B** the head is empty.
Bernard Davis, et al., *Microbiology* (New York: Harper & Row, 1967), p. 1035. Reproduced by permission of Edward Kellenberger, Biology Center, University of Basel.

what one species does to milk to make cheese depends upon the environmental temperature. One cheese may have been discovered in an area with a particular average temperature. This environmental temperature is the *ambient temperature*.

Our major concern in this book is what microbes do, and we find an almost endless variety in nature of different means of survival. For almost every kind of chemical in nature capable of yielding energy, there is a species of bacterium dependent upon that chemical for energy. In Table I, I listed types of microbes and their nutritional characteristics. However, I made no distinctions within each classification. The fungi, including yeasts and algae, offer no problem. The former are all heterotrophs while the latter are all photosynthetic autotrophs. Among the bacteria we find both those nutritional types as well as other types that are restricted to that group of microbes. Specifically, there are bacteria that can "fix" carbon dioxide from the energy derived from the oxidation of inorganic minerals. These species are exceedingly important in the cycling of carbon, nitrogen, and sulfur in nature. (Chapter 7 will treat these in detail.)

It is among the heterotrophs, however, that bacteria exhibit a most remarkable appetite. Some species eat the hydrocarbons in petroleum and some grow in the many sugars found in plant juices and milk. The latter include a most important family, the *Lactobacteriaceae*. These organisms have three things in common: they make acids from sugar; they are fermentative (see Chapter I); and they are indifferent to the presence of molecular oxygen. Some make only one acid, lactic, and are called *homofermentative;* those that make more than one are called *heterofermentative*. These species are involved extensively in the products discussed in Chapters 2 to 4. The ultimate in heterotrophs are the viruses. They can't do anything for themselves. They are extreme examples of obligative parasites. They need all the vitamins, amino acids and other growth factors found *in vivo* (in the host cell) and in the right proportions. Up to very recently, this has meant that the host cells were cultured (cultivated) so that the viruses would have a home. Several investigators have solved that puzzle for two bacteriophages by creating the proper environment *in vitro* (in a test tube).

Although many of the species described later on are active in mixed cultures (more than one species), the science of microbiology is dependent on pure culture isolation and study. Almost all of the processes discussed later on involve pure culture. Cultures of every recognizable species are available; all you have to do to start a microbial garden is to select an inoculum (a starter culture), add it to the appropriate medium, incubate at the optimal temperature (ambient or otherwise) for a convenient period of time and harvest your cells or their products.[10]

1
IN THE BEGINNING

Perhaps it may sound irreverent to use Holy Scripture as a launching pad. Nevertheless, Judeo-Christian heritage and tradition (as well as those of many other religions) are mystically associated with the products of microbiological activity, namely, bread and wine.

Not only are these symbolic parts of two important rituals, the Jewish Sabbath Kiddush and the Christian Holy Communion, they are also referred to allegorically throughout the Bible, for example from Proverbs: "The wicked eat the bread of wickedness and drink the wine of violence." From a purely objective viewpoint, it is not so farfetched to imagine an untutored mind in awe of bread which unaccountably rose from a leavening and doubled in bulk, and amazed by wine powerful enough to turn men's minds, when yesterday it was mere grape juice.

In Genesis 9:21, Noah is mentioned as a planter of the vine: "And Noah the husbandman began and planted a vineyard and he drank of the wine and was drunken." To be sure, this intemperate behavior can be attributed to inexperience. However, the results of his little binge had far-reaching consequences. Of the three sons who discovered him drunk and naked in his tent, only Ham *saw* his father.[1] For this he was cursed forever to be the servant of Shem and Japheth. Ironically, this venerable curse is the basis for apartheid in South Africa. Poor son, to stumble onto a father with a hangover and have his progeny cursed for an eternity!

However, the testaments are replete with advice on the merits and dangers of wine drinking,* and if one message comes through loud and clear, it is continence: "Who hath woe? Who hath sorrow without cause? Who hath redness of eyes? They that tarry *long* at the wine." (Proverbs 23:29) And on the use of wine as medicine and tranquilizers: "Give strong drink unto him that is ready to perish and wine unto those that are of heavy hearts. Let him drink and forget his poverty and remember his misery no more" (Proverbs 31:6–7); "Drink no longer water but use a little wine for thy stomach's sake and thine other infirmities" (I Timothy 5:23); and finally, as a key-note for the rest of this chapter: "Wine that maketh glad the hearts of man" (Psalms 104:15).

How long has man been fermenting the juice of the grape? It has been said that wine is as old as the thirst of man and records of wine-making have been found in the earliest known civilizations. Considering the ease with which fermentation takes place naturally and considering the innate curiosity of *Homo sapiens,* we could assume the earliest fruit and berry eaters had sampled the effects of the "spoiled pick of yesterday." Paleontologists have found fossilized remains of vines in many places in Europe and even in Iceland that bear striking resemblance to the botanical family *Ampeladicea* to which belongs the genus *Vitis.*

I would like to digress a moment and define *wine.* Specifically, *wine* refers only to the fermented juice of the fruit of the vine, the grape, and although any juice or plant extract containing fermentable sugar can become alcoholic, these beverages to the purist are never blessed with the name *wine.* They usually have a very limited clientele, never have "vin-

*Even the French equivocate on this subject. See Figs. 9 and 10.

9. HEALTH . . . SOBRIETY . . .
"There is a grave risk of cirrhossis of the liver when daily consumption of wine exceeds one litre."

10. "Wine is the healthiest and most hygienic of drinks"—Pasteur

tage" (figuratively, "wine age"), and are never extolled as the only *raison d'être* of the fruit itself. Brillat-Savarin, the famous French gourmet, when once offered a dish of grapes by his host, indignantly rose to his full height and said: "Sir, I am unaccustomed to taking my wine in pills." I can conceive of no other beverage being accorded this accolade.

Pictorial depictions of wine-making have been found in the tomb of Phtoh-Hotep from Memphis, Egypt, some 4,000 years B.C., and regulations on the sale and quality of wine were described in the legal code of the Babylonian King Hammurabi, who reigned about 2000 B.C.

In 1967, the oldest known Greek ship was raised from the Mediterranean off Cyprus. This vessel, dating from the fourth century B.C., was apparently engaged extensively in wine traffic, possibly picking up its salubrious cargo on the island of Rhodes. In fact, in some instances, wine trade was a focal point between trading countries.

Perhaps, the outstanding example of this intercourse is the Sherry industry between Spain and Great Britain. As early as the twelfth century, even when the city of origin of Sherry, Jerez, was under abstemious Arab rule, the British went to Spain for their liquid cargo. By the fourteenth century, England had become the world's largest consumer of Sherry. And in spite of wars the British have continued their romance with Sherry through the centuries, consuming in some years 90 percent of all produced.

Let us look at what it takes to make a memorable wine. Although there are many species of the grape vine *Vitis*, only one has produced wines of continuing recognizable quality—that is, *Vitis vinifera*. The wine grape is an incongruity: the most fertile soil may produce the biggest clusters of grapes but the poorest quality of wine. The Latin poet Horace said, "Bacchus loves the hills." *Vitis vinifera* grows best on hillsides (Fig. 11) in soils rather poor in organic nutrients but rich in minerals. This is the land, sometime of volcanic origin, that yields the

11. The 1000-Barrel Mountain. The Wachau on the Danube. The vines are planted on the ranges up the slopes.

wines of bouquet, of body, of breeding, the wines that inspire poets and have created cults of snobs, culminating in that august body, the "Tastevin." Locales zealously guard their registered names (*appellation contrôlée*).

FROM THE GRAPES COMES THE WINE

Essentially, all wines result from the conversion of natural sugars to ethyl alcohol by a species of yeast called *Saccharomyces ellipsoideus*. This hardy species is able to withstand higher concentrations of alcohol than some of its relatives, the so-called wild yeasts. They are selectively killed when the alcoholic content reaches about 4 percent. Only if the sugar content is high enough will the alcohol reach 12 percent, and sometimes higher. Most naturally fermented wines, however, rarely go beyond that level. The amount of sugar in the grape is a function of climatic conditions. Thus sunshine and rainfall play an important part in the growth and maturation of the wine grape and the eventual production of good wine. You read about great wines, chateau bottling, and great years. Obviously, the tender, loving care of the individual *vigneron* (husbandman of the vine) contributes, but beyond that, the location predicts greatness. A rainy, warm spring, a sunny but not too hot summer with adequate rainfall does the rest.

This combination yields the maximum harvest of grapes with the optimal sugar content. Before describing the specific wines found throughout the world and how they are made, I would like to acquaint my readers with some of the general principles of oenology, or wine-making, especially as it applies to microbes.

The metabolic reaction that starts the ball rolling is the conversion of grape sugar or glucose to ethyl alcohol. This series of chemical reactions is referred to as fermentation, called by Louis Pasteur *la vie sans l'air* ("life without air"). This intracellular pathway has been studied intensively since the pioneer work of Pasteur, and, beyond the academic interest in how yeast makes alcohol, has been the prototype for all studies on energy conversion in the cell.

From the lowly yeast we have learned about the need for phosphate in cell nutrition, and how the cell's enzymes store and transform energy. There are approximately eight separate steps in the transformation of sugar to alcohol, and along the way is illustrated the need for the vitamins niacin and thiamin and, except for the last two steps, how muscle tissue converts sugar to lactic acid. Thus, even the most ardent temperance worker can have no quarrel with the coincidental findings on this inconsequential microorganism.

Louis Pasteur was the first to critically examine the conversion of sugar to alcohol. The word *ferment* which so aptly describes this reaction today is used in its original sense: to agitate or stir up. Consider the phenomenon in which a solution of sugar slowly turns to alcohol and at the same time gives off carbon dioxide, causing the vat to bubble like the witch's cauldron. No wonder agitation is synonymous with fermentation! In 1838, Pasteur presented the first scientific evidence to the world showing that budding globules increased in number during the alcoholic fermentation, thus establishing a cause-and-effect relationship between growth of yeasts and production of alcohol. In defining the requirements of fermentation, Pasteur discovered one of the keystones in cellular carbohydrate metabolism, namely that air or oxygen added to a fermenting system turns a biochemical switch resulting in a bypass of alcohol formation and leading totally to carbon dioxide and water. This reaction to oxygen is now called the "Pasteur Effect" and is found in all so-called facultative organisms. These are organisms that are ambivalent with respect to molecular oxygen or oxygen as in air. They can live in its absence, in which case they ferment available substrates (food) or they preferentially *respire* when given the oxygen. This metabolic schizophrenia has evolutionary advantages, since it allows certain species to exist literally in two worlds. However, as far as the cell is concerned, fermentation is a very poor source of energy compared to respiration: it gets only one-twentieth the amount available from sugar with oxygen, and the alcohol is a dead end for the yeast; and indeed, when sugar is not growth-limiting (in excess), the levels of alcohol produced do result in their death. Since 95 percent of the sugar calories are still in the alcohol molecule, it is easy to appreciate why the "drinking man's diet" meets so rarely with success—a doughnut and a daiquiri have about the same caloric value.

Although the major ingredient formed during yeast fermentation of sugar is ethyl alcohol, the resulting infinite variety of wines is due to a large number of other products, both found naturally in the grape juice and formed by the yeast cells as they metabolize. The major acid in grape juice is tartaric acid, which apparently protects new wine from bacterial spoilage and eventually crystallizes out as "cream of tartar" and falls to the bottom of the barrel in the lees. About 95 percent of the sugar is converted to ethyl alcohol and carbon dioxide. The remaining 5 percent is used by the yeast for growth and is converted to glycerin and a miscellaneous array of organic compounds, such as alcohols other than ethyl, acids, and aldehydes.

Perhaps of these, succinic acid adds the most for its small presence (0.6 percent of the sugar fermented). Pasteur attributed the wine flavor to this compound. The other alcohols and acids, over time, form a chemical combination called an ester, and its volatility gives each wine its distinctive aroma and bouquet.

You can see readily why it is impossible then to call a California red wine a Burgundy despite the fact that *Saccharomyces ellipsoideus* is the yeast and Pinot Noir is the variety of *Vitis vinifera*. The soil and climatic conditions determine the final content of the grape and, therefore, the nature of the wine.

A practical result of the "Pasteur Effect" was soon learned by novice winemakers and brewers. Excess air during fermentation gives more yeast and no alcohol. In addition, there were nuisance microorganisms (which this book is dedicated to avoid) that converted alcohol to vinegar (this is an example of a reaction that in a different place and time is desirable; it will be discussed in that context in Chapter IV).

Although *Saccharomyces cerevisiae var. ellipsoideus* is the dominant species involved in the alcoholic fermentation of must (the technical name for fermentable natural grape juice) a variety of other yeasts can contribute to the final product. The distribution of at least twenty other species from a number of locations from the plant parts to wine press, barrel, and bung and bottling operation, as well as during fermentation, have been reported all over the wine-producing world. Despite the prevalence of some species in certain wines in different locales alone, the possibility that certain yeasts may be particularly important, for example in flavor formation in Sherry, does exist.

There is no hard opinion on the value of mixed yeast as opposed to pure culture of *ellipsoideus* in producing a most acceptable finished product. The reader is referred to a recent definitive review on the microbiology of wine-making for more details.[2] Since there is no proof of the advantage of a so-called pure culture, no yeasts are added nor are any necessarily destroyed. I would note for your interest that in beer-making, the natural yeasts are all killed, and fermentation is carried out by the single strain of brewer's yeast added to the wort (pre-beer).

The beginning of wine-making is the collecting of the grapes. Hopefully, the harvest will yield a memorable vintage. The vine has many natural enemies, and a century ago the great wine-producing districts of Europe were all but wiped out by a plant louse called *Phylloxera vestatrix*. The eventual salvation of the vine was the California cousin of *Vitis vinifera, Vitis rupestris*. The roots of the latter were resistant to the louse, and so the leaves and fruit-bearing parts of *Vitis vinifera* which fortuitously resisted the louse could be grafted to the roots of *rupestris*.

Today, all the wine-producing vines are the results of that fortunate horticultural union. In addition to insects, the *vigneron* must also worry about mildew from certain fungi. These have proven treatable with chemical agents. One of these, Bordeaux mixture (copper sulfate and lime), was first introduced to the Bordeaux district of France and since has achieved general distinction as an agricultural fungicide. However, not all fungi are enemies of the vine; for certain white wines in which sugar content of the grape is very critical to the development of appropriate bouquet, growth of *Botrytis cinerea* (Fig. 12), a grey powdery mold on the grape, is desirable. This mildew literally sucks water out of the sap of the fruit, thereby increasing the relative sugar concentration. Because of the esteem the French and the Germans place on this accidental fungal infection, they refer to it respectively as *pourriture noble* and *Edelfaule*: the noble mold!

12. Mildew on the Grapes (right side)
Reproduced, by permission, from Andre Simon, *Wines of the World* (London: George Rainbird Ltd., 1967), p. 354. Photo by Percy Hennell, courtesy of George Rainbird, Ltd.

Once the grapes have been collected, their fate depends upon the general type of wine being made. We can classify wines in many ways: by their color—red, pink, or white; by their carbon dioxide content—still or sparkling; by their alcoholic content—natural or fortified; by their residual sugar content—sweet or dry. In addition, the vineyard, district, or country superimposes its own parochial characteristics on wines that evolve from the peculiarities of geography and the mystique of the local *vigneron*.

STILL WINES: RED, WHITE AND PINK

To make red wine dark, blue-black grapes are used, and the total *vendange* (harvest) of stems, skins, pits, etc., is pressed together so that the must contains these parts in addition to grape juice. The dark red color comes from pigment cells just beneath the skin which are ruptured when the grapes are crushed. White wine can be made from either white or black grapes, but only the juice of the grape is used, care being taken not to release any pigment if the latter grapes are selected. Pink or rosé wine is made from dark grapes in which slight pressure on the skins produces the desired color. The method of pressing varies from the traditional stamping by foot to the use of huge mechanical presses.

The presence of stems and seeds gives red wine its very distinctive "dryness" or, to the less sophisticated palates, "sourness." This is due to the extraction of tannins which, in addition to imparting a flavor, actually protect the wine from spoilage. Because of all the extractives, red wine is a much more complicated microcosm of events than white wine. Even long after fermentation is complete, red wines continue to mature. One bottle of Bordeaux red wine from one of its most famous vineyards, Château LaFite, was bottled in 1811 and was still exceptionally palatable when opened 140 years later. White wine, on the contrary, is at its peak very soon after bottling and, if anything, begins to deteriorate in quality with excess aging. Thus, it hardly pays to keep good white wine beyond one season.

Since the must is already inoculated (the surface of the grape is covered with yeasts), fermentation begins immediately. Successful fermentation is very temperature-dependent: above 90°F and below 70°F activity is minimal; yeasts prefer 85°F. If ambient temperatures fall outside of this range, it is necessary to take corrective measures. This is meant as a helpful hint to would-be wine-makers.

During the fermentation, the must is kept agitated by the constant release of carbon dioxide. This is a fortuitous event, since the bottom deposit, or lees, has to be continually mixed. The presence of a carbon dioxide atmosphere serves the dual purpose of stimulating fermentation and excluding oxygen in the immediate atmosphere above the fermenting surface. This result prevents the growth of the vinegar bacterium mentioned earlier. It is convenient to place a loose-fitting top over the vat to contain the CO_2 without permitting the buildup of excessive pressure. When most of the available sugar is used up, the wine is transferred to casks or barrels. The size and wood depend upon the wine.

Here fermentation is completed—sometimes because the alcoholic content kills the yeast; or because the sugar is used up; or, in the case of some white wines, the process is stopped forcibly by heating (pasteurization). In the casks, the dead yeast cells and other debris settle to the bottom, and external agents such as egg white or casein are added to clarify the wine. After a suitable period, the wine is bottled.

However, during the period in the cask, great care must be taken to prevent the entry of air. With time, there is evaporation through the pores of the wooden staves. This liquid is replaced by air. If the air space is not itself replaced by more wine, the risk of vinegar production is very great. To facilitate clarification, the wine is transferred several times to fresh casks, care being taken not to stir up the bottom sediments. This process is known as *racking*. The treatment of a still wine during this period is directly related to its pedigree. The *vin ordinaire, vino común,* or *heurige Wein* is treated much more casually. The alcoholic content of these wines varies from 10 to 13 percent, the final value dependent upon the sugar content of the grape and whether additional sugar was added. Also, the *ellipsoideus* yeast has an upper tolerance of 14 percent, and under most conditions naturally fermented wines rarely exceed that value.

SPARKLING WINES

Essentially, sparkling wines are nothing more than the still wines described above in which the fermentation is completed in the bottle. With appropriate pressure-resistant stoppers, the carbon dioxide released is entrapped in the wine. Actually there are many more refinements, and great care is necessary at each step to make a product worthy of all the individual effort.

The undisputed monarch of all sparkling wines is Champagne; all others are imitators. Nevertheless, the process is the same. Sad to say, the appellation *Champagne* has been used so cavalierly outside of its country of origin that its true significance is lost on most people. In France, Champagne must be made from the Pinot grape, either white or black, grown in the area around Rheims east of the Marne River, the Département Marne. This is Champagne country. Only the best of the harvest is used, and the final fermentation must be in the bottle. If you see a Champagne label that says "bulk fermentation" (something you will not see in France), you know that the bubbles were added after the wine was finished. This type of carbonated wine (*not* Champagne) is always much less expensive.

Since Champagne is partially fermented in the bottle, certain problems are created that are not encountered in still wine. To begin with, only must from the first pressing, or *cuvée,* is used. The best Champagne grapes result from very hot summers punctuated by violent thunderstorms, an almost impressionistic forerunner of the personality of this delightful beverage. The *cuvée* is placed in a 450-gallon vat where it ferments violently for about thirty-six hours. This boisterous bubbling brings the lighter scum to the surface while heavier debris (lees) settles to the bottom.

The pre-Champagne is transferred to ten oak casks, and fermentation is allowed to continue until all the sugar is gone, while allowing the carbon dioxide to escape. This step takes about eight to ten weeks. These casks are then racked several times during a four-to-five-week period after which the *chef de caves* blends the various wines prior to bottling. Since fermentation is to continue after bottling, additional sugar has to be added. It is rock candy dissolved in Champagne wine (*liqueur de tirage*). This step is primarily to add the sparkling quality, carbon dioxide. The filled bottles are corked, clamped shut, and placed on their sides in cool cellars to complete fermentation and maturation.

They are left as long as two to three years, and during that time objectionable sediments form. The art is in removing these withouth allowing wine or carbonation to escape. This is done by daily twisting of the bottle and gradually bringing them from the horizontal to the vertical position so that the sediment slides toward the neck and eventually accumulates on the cork. In the old days the responsible individual (*le dégorgeur*) would quickly pull the cork, place his finger in the neck, and put in a new cork. Today this process is facilitated by freezing the neck with dry ice, allowing the cork and sediment to be removed without loss of contents.

The final step in finishing the Champagne is the addition of more sugar, this time dissolved in brandy and wine for the purpose of making the various degrees of sweetness or dryness in your final glass: *brut, extra sec,* and *demi-sec*. After the sugar is added, the cork is forced in and wired to the neck. Perhaps, you now can better appreciate why Champagne costs as much as it does.

There is a current rage for one carbonated wine called Cold Duck. Let me give you some folklore on its origins. In the last century in Bavaria it was customary for early morning hunters to take with them bottles of sparkling Burgundy to warm them as they pursued the boar. They returned with many started flat bottles of sparkling Burgundy. Being very frugal, they revitalized the contents of those bottles with fresh Champagne. They called this mixture Kalte Ende, literally, "the cold end." Somehow this has been bastardized into Kalte Ente, literally, "the cold duck."

FORTIFIED WINES

These are wines that reach the limits of natural fermentation and presumably are increased in alcoholic content by the addition of brandy derived from the distillation of that wine.

There is considerable controversy over what constitutes the upper limits of natural fermentation. Values of 14 percent for *Saccharomyces cerevisiae var. ellipsoideus* to 19 percent for some wild species have been reported. Nevertheless, it is common practice to add alcohol (usually distilled from the same wine) to wines above 14 percent.

Although there are numerous regional fortified wines, those that have acheived universal distinction all come from approximately the same part of the world: the Iberian Peninsula and the Madeira Islands. These wines, much imitated and even more abused in imitation, reflect the maritime influence of the British Empire, for without the interest of the English, Sherry, Port, and Madeira would never have achieved their present level of acceptability and prominence.

Of the three, Sherry is the most copied and consumed, and offers perhaps the greatest varieties. However, the making of fine Sherry is most challenging and exciting, since the end result is so unpredictable, as I will explain later. The place of origin of Sherry centers in a city in the south of Spain, Jerez de la Frontera. Jerez was known to the Greeks as Xeres and as

Sherich by the Moors. In 1204, the city was taken by Alfonso X of Spain from the Moors and, at that time, was dubbed Jerez de la Frontera, since it bordered on the Moorish Caliphate of Granada. This latter part of Iberia did not join the Spanish Kingdom until 1492 when Ferdinand and Isabella completed the Christian domination of Spain. However, the name "de la Frontera" has stuck to this day.

The soil that yields the finest wine of the district is called *albariza* (chalky), and although there are many varieties of vine grown in Andalusia, just a few are suitable for producing good wine. Primarily, these wines come from the Palomino grape and, for the sweetest, the Pedro Ximenez. These grapes are harvested in mid-fall, and traditionally, the *mosto* (first pressing) comes from foot-stamping with nail-studded shoes. This is done in *lagares,* shallow, twelve-foot-square bins, two feet deep, with sloping floors with holes that allow the juice to flow through. Usually, this is done at night so that the heat of the day (and it gets very hot in this part of Spain) does not initiate premature fermentation. The *mosto* is put immediately into special oak butts (casks) and left in the sun where primary fermentation begins. Primary and secondary fermentation take two to three months, and it is the latter which is so typical of Sherry.

In contrast to the still wines described earlier, the presence of oxygen is necessary to the production of Sherry. Therefore, air space is left in the butt after filling, and the bung is stoppered with a loose-fitting cork. The oak staves are sufficiently porous to allow a certain amount of evaporation during the process. Secondary fermentation involves the growth of a surface film of yeast called the *flor* (flower) without which the wine cannot become Sherry (Fig. 13). The butts, containing about one hundred gallons each, are stored in tiers in huge buildings of cathedral-like dimensions that permit maximum light and circulation of air.

And now, the caprice of Sherry! Even though all the wine may derive from the same vintage, what results is a variety

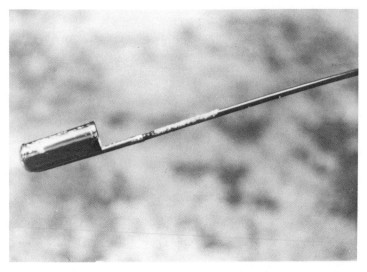

13. Secondary Fermentation Involving Growth of a Surface Film of Yeast: *Flor*
Flor is white film on dipper.

that depends on human expertise for classification. The young wine is called *vino de añada*—literally, new wine or wine for addition—and is placed in two general categories: *fino* and *oloroso*. The former are paler in color, very dry, and delicate, while the latter approach amber, have less pungency but, nevertheless, have a very distinctive bouquet. Even though the sugar is completely gone, they tend to have a slightly sweet or winey aftertaste. Both of these basic groups are further subdivided; for example, *amontillado* is more pungent than *fino* but less sweet than *oloroso*. These various categories are ready for the next steps in the maturation process which makes Sherry so unique.

    Solera Process—*Vino de añada* (literally wine of addition) appropriately labeled and fortified with Sherry brandy, is placed in butts in the *bodega* (cellars) called a *criadera* or nursery. This wine serves as a reservoir for feeding the *solera*, a many-tiered (as many as ten) pyramid of butts in which the bottom tier represents the year of origin of the *solera*. The top layer is the youngest wine and is replenished from the *criadera* (Fig. 14). Wine for bottling is drawn from the bottom which is continually being blended by sequential additions of generations from above. Thus for every *fino, amontillado,* or *oloroso* determined as *vino de añada,* there is a separate *solera,* and the mixing of similar wines from vintages from many harvests assures a finished product of uniform excellence.

Theoretically, by this process Sherry always contains a fraction of very mature wine. However, the labeling of a Sherry "Solera 1895" means only that the *solera* was laid down in that year and not that the wine is that ancient. The sweetest of Sherries, the so-called creams or milks (an English description not used by the Spanish) are made from the Pedro Ximenez *mosto* and are used straight or for blending with *oloroso* to give the desired level of sweetness.

The marketing of Sherry, like that of Port and Madeira, reflects the English interest in that trade. Much of the selling of Sherry around the world is by vintners who are English-derived such as Harvey and Sandeman, or by partnerships like Gonzalez, Byass (Fig. 15), or Humbert and Williams, in which the first-named is the resident Spaniard who runs the production end, and the last is the Englishman, who sees to the distribution. The brands I have named probably represent a large proportion of all the Sherry produced, since not only do they bottle under their own label, but also they sell in bulk to others who do no more than blend and bottle. Perhaps the only major *solera* without British connections is that of Pedro Domecq. For those of you who will visit Spain, visit Jerez and accept the hospitality of one of large wineries. You will find it an unforgettable experience.

ON THE SELECTION AND DRINKING OF WINES

Sad to say, the drinking of wine in many areas has developed into a cult where attention to the color of wine, the temperature at which it is served, and the kind of glass tends to be overre-

14. The Criadera
The beginning of the *solera*.

15. Partnership of Gonzalez, Byass
Pouring the Sherry.

strictive. I am sure many people shy away from serving or ordering wine for fear of making a mistake. There are many truisms on wine-drinking! White wine—chilled with fish; red wine—cellar temperature with meat; etc. But if you chill red wine and like it, no strange curse will fall on your head. One of the most delightful of summer drinks is Sangría from Spain, made from red wine, ice, and some sugar and citrus juice! Wine-drinking should be an adventure, and the proof is in the drinking! *In vino veritas.*

2
GIVE US THIS DAY
OUR DAILY BREAD
(AND BEER)

I recall seeing a television play once in which a blindfolded kidnap victim was able to lead the police back to the hideout where he was held by taking note of the sounds and smells along the way. Just recently, I returned to New York City for a short visit, and as I drove from La Guardia Airport I passed my old neighborhood in the South Bronx; even on the freeway and without looking, I knew I was there. The smells of fresh bread from the Ward Baking Company flooded my olfactory sense. What a delightful aroma, and what nostalgia it evoked; what pleasant memories it conjured up! Who among my reading audience does not remember with similar delight the pungent aroma from the neighborhood bakery or his mother's kitchen. How sad that most places where we now buy our bread are so sterile!

One of the attractions of most cities in Europe is that their citizens still worship the taste, the feel, and the aroma of freshly baked bread. The Frenchman in particular is a zealot regarding his bread. Not only must it be fresh (baked several times daily), but also it must have a specific shape for each meal and occasion! A classic French film of the late thirties, *La Femme du Boulanger* ("The Baker's Wife"), describes the utter collapse of a small village when the baker stops baking, following his wife's desertion.

Even in 1970, the intensity of feeling for bread continues in France. The central government in Paris fixed a price on a standard loaf at 94 *centimes* (Fig. 16). This created a demand for one-*centime* coins that were in short supply. The bakers rebelled, increased the size of the loaf, and raised the price accordingly. The government threatened the bakers with fines and penalties, forcing them to revert to the legal loaf. One-*centime* pieces will now be minted at a cost exceeding their face value. The Frenchman is not the exception—he is merely the best example of how "civilized man" regards his loaf.

Bread is more than just food. It is symbolic of all of man's physical needs. Perhaps a more striking example of the significance of bread can be summed up by the response of an over-fifty taxi driver in Madrid when I asked him what he thought of an Army Day parade that had taken place the previous afternoon in the spring of 1965: "Menos desfiles, más pan blanco" ("Fewer parades, more white bread"). "Dough" (prebread) has been slang for money for many generations: today the euphemism has been dropped, and the with-it generation just calls it "bread." The origins of bread are buried in antiquity, but there is no doubt that it is one of the oldest prepared foods known to man. Christians and Jews and so-called heathens have sanctified it, and poets have praised it.

Although the variety of ingredients that can be included in recipes for breads from all over the world is extensive, the essential requisites need only be flour, water, and a source of leavening. Undoubtedly, primitive man went through many generations of eating the unleavened progenitors of bread,[1] and when he stumbled on the advantages and sources of leavening is certainly unknown to us; however, beer-making or brewing seems to have been practiced at least six-thousand

16. Standard French Bread Loaf

years ago, and there is pictorial evidence from an early Egyptian dynasty showing a brewery adjacent to a bakery, suggesting that these ancients were aware of the relationships of the ferments of beer to the rising of bread. To this day, *baker's yeast* and *brewer's yeast* are practically synonymous, and indeed some of the largest suppliers of yeasts to the bakery are the breweries. Thus in this chapter the two foods will be treated with emphasis on their historical and technological kinship.

In Figures 17, 18, and 19 are shown the leavening and baking rooms of a typical French bakery, and in Figure 21 a flow diagram for beer manufacture in a large American brewery. In the leavening room and in the fermenting vats, the same biological activities are being carried out by yeast.

THE LEAVEN—PAST AND PRESENT

The Hebrews were a nomadic shepherd people who came out of the desert to accept the agricultural heritage of the Babylonians. With this heritage came the tradition of a complete cleaning of the granary with the advent of the spring harvest.

17. Leavening Room

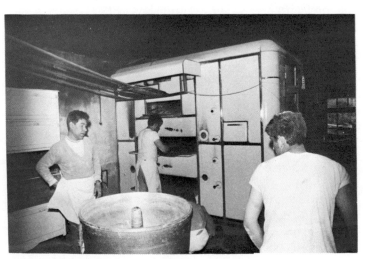

18. Baking Room of a Typical French Bakery

19. Out of the Oven

This also included getting rid of any fermented grain which in ancient times served as one major source of leavening. Thus the first bread of the spring was unleavened. This ritual was adopted by the Hebrews and combined with their spring pastoral celebration of the sacrifice of the young lamb. Both the eating of the unleavened bread and the "paschal lamb" have become symbolic of redemption for Jews and Christians, albeit interpreted differently than originally.

Although the absolute essentials for bread are flour, water, and leavening, the results one gets are very dependent on even the nature of these three ingredients. Let us look at the leavening first to see what it does, and how it does it.

Saccharomyces cerevisiae is a true yeast, a kissing cousin of the *Saccharomyces ellipsoideus* whom you met in Chapter 1. It is the species used in purposeful inoculation of the flour-water dough or sponge. Under suitable conditions the small amount of natural sugar in the grain is fermented to CO_2 and ethyl alcohol. Essentially, this is a process of resting metabolism, since during the rising or inoculation period there is little or no multiplication of the yeast. In fact, at a starting level of 2 percent of the total sponge, the yeast population increases by 50 percent; then it levels off. Afterward, only the substrate is metabolized without an increase in population. The two products of fermentation have different destinies. The CO_2 becomes trapped in the sponge (this is dependent on a special quality of the flour, which will be discussed later) and causes it to rise. During baking, these gas pockets are fixed as the dough is cooked, giving the bread its spongy texture. Although the center of the dough rarely exceeds 212°F, this temperature is sufficient to vaporize all of the alcohol produced. The doughs made with yeast are commonly referred to as sweet, and produce a variety of breads, depending upon the origin of the flour and the temperature of leavening.

Perhaps a more primitive and less predictable or controllable method of leavening involves the use of naturally fermented dough maintained as a continuous culture. Usually, the microbes naturally on the grain are allowed to ferment in the sponge. This type of reaction produces a sour dough due primarily to the growth of so-called heterofermentative lactic acid bacteria. The major metabolic products are acids like lactic and acetic, as well as carbonic. Several of these organisms belong to a genus (group) known as *Lactobacillus* (*L. plantarum, L. brevis,* and also *Leuconostoc mesenteroides*). These organisms prefer average room temperature (77°F). At higher temperatures (90 to 95°F), other species would grow whose by-products are less favorably tolerated by the palate.

Although this process is perhaps more usually found in the baking of sour rye bread, it certainly is not restricted to that flour (indeed, San Francisco sourdough bread and the appellation "sourdough" applied to the Klondike prospectors refer to a wheat-derived sponge).[2]

However, for most commercially available bread, *S. cerevisiae* alone or as an adjunct to sour sponge is still the prime source of biological leavening for baked goods. The origins of the yeast used in bread, although the same species as that used for beer, is rarely taken from the brewing process. The presence of residual hops and the mashing and cooking together would add flavor and color impossible to remove. Thus, a bread made from beer-produced yeast would be darker than most, and bitter. In addition, when compared with distiller's yeast, the beer-produced yeast has poorer leavening qualities, nor is it capable of relatively fast fermentation at high temperatures.

Yeast as a by-product of grain alcohol production is acceptable for baking. In addition, baker's yeast is produced on its own by growing pure cultures of the proper strain of *S. cerevisiae* in a nutrient solution based on molasses or corn sugar and certain B-vitamins. In contrast to brewing, the large culture tanks are actively aerated for maximum growth rates. The yeast is harvested, washed, and compressed. A fermentation vat of 15,000 gallons can produce one ton of compressed yeast per day. Fresh yeast can be kept up to three months at refrigeration temperatures (45°F) without appreciable biological loss. Dehydrated yeasts can be kept longer but are never as active as fresh cells.

Equal if not greater in importance to the leavening is the flour. Not all grains yield flours that leaven. It is no surprise then that wheat is the major grain crop for the making of bread. It contains the right combination of natural contituents to result in the maximal rising of the dough.

Practically all the flour used in baked goods comes from the botanical species *Triticum vulgare,* while the other major wheat species, *Triticum durum,* is the source of semolina flour used in noodles and macaroni. *T. vulgare* is divided into horticultural varieties based on time of growth and harvest (winter and spring wheat) and on color (white, red, and yellow). Different uses require different grain properties, and for bread-baking the most desirable is a so-called strong flour with a relatively high gluten content. Gluten is a protein giving elastic properties to dough. The flour is derived from the very center of the wheat grain (Fig. 20) and constitutes 82½ percent of its total. The outer four layers of the grain make up about 7 percent and yield what is called bran. Between the endosperm and the bran is the wheat germ, which has its own special uses. Sometimes the entire grain is used for whole-wheat flour. However, our main interest is in the more generally used endosperm product from which flours are made.

The average constituents of wheat flour are gluten-forming protein (11 to 17 percent), starch (69 percent), sugar (2.5 percent), water (15 percent), and miscellaneous fat, protein, minerals, and vitamins. High gluten content is most desirable, and as with the wine grape, soil and climate play a major role in determining the quality of the harvested product—the most favorable being deep, rich soil, moderate rainfall, snowy and severe winters, and sunny and pleasant summers.

Although the separation, grinding, and milling of flour have some relevancy to the final product, such a discussion is not really germane to the major intent of the book and I invite my reader to seek this information in the list of suggested readings provided. However, I would like to mention that flour manufacture in times past helped make the rustic landscape that much more picturesque—dotting the countryside with windmills and watermills. One seldom sees either of these today, all milling being done in sterile factories.

The kinds of breads and rolls that are available are almost infinite, and this variety not only reflects individual tastes and preferences of a region but also is a function of the type of flour that predominates and the prevailing ambient temperature forced on the baker for leavening conditions. Although the requisites for leavening are simple (e.g., a small amount of fermentable sugar, like maltose), other ingredients can stimulate yeast activity, as salt does, while others can retard it, as do certain oils and flavoring agents. To achieve a better appearance and keeping quality, natural flours are chemically

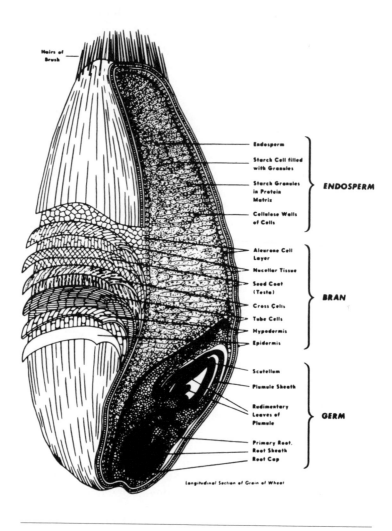

Hairs of Brush

Endosperm

Starch Cell filled with Granules

Starch Granules in Protein Matrix

Cellulose Walls of Cells

ENDOSPERM

Aleurone Cell Layer

Nucellar Tissue

Seed Coat (Testa)

Cross Cells

Tube Cells

Hypodermis

Epidermis

BRAN

Scutellum

Plumule Sheath

Rudimentary Leaves of Plumule

Primary Root.

Root Sheath

Root Cap

GERM

Longitudinal Section of Grain of Wheat

20. Sketch of a Wheat Grain
Reprinted, by permission, from Edmund Bennion, *Breadmaking: Its Principles and Practice,* 4th ed. (London: Oxford University Press, 1967), p. 12.

bleached. This undoubtedly affects flavor and nutritional value, but has no effect on subsequent leavening activity. Also, in recent years commercially prepared breads have as an ingredient an antimold agent such as calcium propionate. This too apparently has no effect on yeast action.

What specifically is expected of yeasts and other microbes during the baking process? There are three main functions:

(1) to produce carbon dioxide in sufficient quantities and at the right time to inflate the dough and produce a light spongy texture, resulting in a palatable bread when correctly baked;

(2) to produce a complex mixture of many types of chemical compounds which contribute to the flavor of the bread;

(3) to help bring about the essential changes in the gluten structure, known as ripening or maturing of the dough.

The general mechanism for sugar fermentation by yeast has already been discussed in Chapter I: yeast can ferment natural sucrose or maltose in flour to glucose, and then to the final CO_2 and ethyl alcohol; however, it is unable to digest starch. Further on in this chapter I will discuss how the brewer gets around this problem.

Yeast cells also produce a large number of intermediary by-products (e.g., glycerin, succinic acid)—the variety depending on the type of flour and the ancillary ingredients. Perhaps as important as CO_2 and flavor is the production of protein-digesting enzymes by the yeast cells that partially digest the gluten and give the dough its spongy quality. Overfermentation can destroy this quality, which is responsible for CO_2 retention. Oftentimes, with very high gluten flour, the proteinase activity of yeast is insufficient for conditioning. It is then necessary to add an external source of enzyme.

Thus, wheat flour, because of its unique properties, is the major grain in breads and cakes throughout the world, whether it is Vienna bread, which is baked in a high humidity oven to develop its flaky crust, or those two all-American favorites: bagels, boiled in water before baking; or pizza dough, with its highly elastic properties.

Other grains are used also, although to a limited extent. The only significant competitor to wheat is rye flour. The internal structure of the rye grain is quite similar to that of wheat, as are the qualitative constituents. The flour is a darker color; even after bleaching, which improves its baking qualities, it has a rather greyish appearance. Although its gluten proteins are similar to those of wheat, it is difficult to get a sponge of any volume from rye—the dough tends to be more closely textured. Because of the sharpness of taste and the poor leavening qualities rye is usually mixed with wheat flour in varying proportions, all the way from one of wheat and three of rye, to the converse.

Perhaps the major use of rye flour is for the making of sour rye bread. In this case the rye flour-and-water sponge is allowed to ferment naturally. The lactic acid bacteria and other naturally occuring microbes convert sugars and nitrogenous substrates to acids and gas (CO_2 primarily); this sour sponge is then subsequently used as a leavening for the "sour" bread. Depending upon the level of sourness and ryeness, the baker can use more or less of this sponge and more or less wheat flour as adjunct; yeast can be used as additional leavening if desired.

Another major bread derived from rye flour is pumpernickel.[3] Traditionally, this bread has been made from unbolted rye flour, like German *Schwarzbröt* (black bread). Today, most of what is called pumpernickel is just a darker cousin of many of the varieties of rye bread mentioned above, darkened, no doubt, by molasses or caramel coloring.

The other major grains of the world, such as oats, rice, maize, and barley, lack gluten and, therefore, cannot be used in leavened bread, except in combination with wheat. They do have other uses, and rice, maize, and barley will be discussed in the second half of this chapter.

THE UNIVERSAL BREW

Beer is derived from the Anglo-Saxon word for barley, *baere*, and is an indication of the importance of this grain to beer. Since species of *Saccaromyces* are unable to enzymatically digest starch to get the sugar to make ethyl alcohol, it must be predigested for them. This is the prime function of the barley. Germinating sprouts of barley (*Hordeum vulgare*) produce an abundance of the enzyme *diastase,* which converts stored starch to sugar for energy. Since the end product of diastase activity is the sugar *maltose* (two molecules of glucose linking together), the germinating process is called *malting,* and resulting enzyme preparation is called *malt.*

In addition to serving as a source of starch-digesting enzyme, the barley grains also provide starch as a fermentable substrate. The proportion of malt- or barley-derived starch to total carbohydrate from other sources such as rice or maize is dependent on the kind of finished product that is desired. Beers derived from 100 percent barley grain are sometimes referred to as malt liquors and are the major types manufactured in beer-producing countries in Europe.

Quite often, the brewery buys its malt from special suppliers in dry form. The brewmaster then mixes this malt with water and maintains a temperature of 105°F for about thirty minutes. This begins the breakdown of barley carbohydrates. The malting process is allowed to proceed only long enough for the sprout to fully synthesize its complement of enzymes. Before autodigestion begins, germination is stopped by slightly raising the temperature of the steeping suspension, care being taken to keep the heat below the inactivation level for the enzymes.

In most cases, the malt carbohydrate is augmented by steeped ground corn or rice which is called *adjunct grain.* The mixture is heated gradually, activating and releasing numerous starch- and protein-digesting enzymes. This cooking process is called *mashing,* and the rate of heating is extremely critical, since enzymes have specific activation temperatures. For ex-

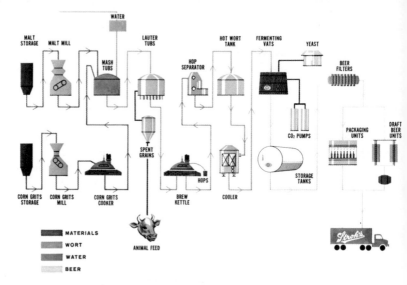

21. Flow Diagram of Beer Manufacture in a Large American Brewery
Reproduced, by permission; from Brew House, "Brewing Process
Flow Diagram" of the Stroh Brewery Company of Michigan.

ample, if protein-digesting enzymes are not maximally active,
residual protein can cause clouding of beer when chilled; on
the other hand, complete loss of protein can destroy its head-
holding ability.

Starch digestion is also critical. Although glucose is
the only molecular species found in starch, it is not just a
simple carbohydrate made of a large number of glucose molecules. Instead, there are two fractions, one called a *straight
chain* (amylose) and another referred to as a *branched chain*
(amylopectin). As you might expect, two different enzymes are
involved. Amylose is attacked by B-amylase which chops off
glucose molecules two at a time; this packet of two is called
maltose. The enzyme is most active around 140°F. Amylopectin
branches are broken into smaller straight pieces by α-amylase
which is most active between 150 to 170°F. These short pieces
are called *dextrins* and are not fermentable, but do contribute to
foam stability. Thus, careful control of mashing temperature is a
major factor in what will become the finished beer.

The solution of amino acids, peptides, proteins, sugars, dextrins, etc., that results from mashing is called *wort* (German for "pre-beer"). At this stage, the wort is transferred to a *lauter* tub (from the German for "clearing") that permits the separation of the spent grain from the wort. The grain is washed with hot water and these washings are added to the wort. The spent grains are sold as cattle feed to yield contented cows.

The clean wort is boiled to completely inactivate its enzymes, and then hops are added—more specifically, the dried flower of the female plant of *Humulus lupulus,* 1/2 to 1 1/2 pounds per barrel of beer depending on the taste desired. Several properties are attributed to hops: certainly a slightly bitter taste, and possibly some of its resins and oils are slightly antiseptic and help prevent deterioration of the finished product. Also on the plus side are the tannins leached from the hops; these combine with the wort proteins that precipitate out of solution, and thus prevent potential clouding. The wort has now reached the stage of fermentation.

In general, there are two basic types, referred to as "top" and "bottom" fermentation, since, in the former, the yeasts rise to the top and, in the latter, they fall to the bottom of the tank. The products they yield are quite different, as are the strains of yeast, their temperature of fermentation, and, in most instances, the amount of malt. Top fermenting yeasts are varieties of *Saccharomyces cerevisiae* and are used primarily in English-type brews such as porter, ale, and stout. These tend to be dark, poorly carbonated, and slightly higher in alcoholic content than bottom-fermented brew.

This latter type is originated in Germany and is called *Pilsener* or *Lager.* The yeasts used in these brews are varieties of *Saccharomyces carlbergensis,* originally isolated in the famous Carlsberg Breweries in Copenhagen.* Pilsener is fermented at

---

*An interesting historical note: Jacobsen, the founder of the brewery, in 1845 built it on a small hill and named the brewery after his son Carl; thus "Carl's hill" or "berg."

about 4°C for 10 days, the resultant beer having an alcoholic content of 3.5 percent; English ale is fermented at about 11°C for a shorter time to an alcoholic level of approximately 4.5 percent.

After fermentation, the brews are aged at 0°C for about a month during which time they mature and yeasts settle out. The latter process is called *lagering* (from the German for "storing"). The lager is then passed through filter beds of diatomaceous earth that remove most of the suspended yeast cells. Beyond this point carbon dioxide is added back, either artificially or by a secondary fermentation (German *Krausening,* "bubbles").

Traditional draft beer is barreled, refrigerated, and consumed as soon as possible. Even though filtered, the brew still contains many microorganisms capable of spoiling the finished product; thus, its shelf life is quite limited. In order to increase its longevity, bottled or canned beer or ale is heat-pasteurized to 140°F for up to fifty-five minutes. This effectively destroys any potential spoilage organisms capable of growth in the beer. (I should emphasize here that organisms of human disease are never a problem, since the acidity of beer precludes their growth.)

There have been recent innovations, enabling shelf-stable draft beer in bottles! After the prefiltration described above, beer may have as few as three living microbes per fluid ounce. These can be removed by a finishing filtration with membrane filters immediately prior to bottling. The sealed unit is biologically sterile for all practical purposes and will not spoil from microbial action. However, purists refrain from calling this draft beer, even though it is not heat-pasteurized. The reasons are straightforward. Beer may not deteriorate from microbes, but it is a fact of life that beyond one month beer does not improve with age, and chemical changes that are time-, temperature-, and light-dependent do produce undesirable effects. True draft beer is kept in the dark, refrigerated, and drunk within several weeks. It is this product that delights the pedigreed palate.

A word about a traditional beer: in the month of March, close to the beginning of the Spring equinox, it has become customary to drink what is called *Bock* beer. It is almost apocryphal that this very dark lager represents the dregs of the brewery accumulated over winter, and that its festive consumption previewed the arrival of a new season's beer. *Bock* is German for ram, the sign of Aries, which begins March fifteenth. The emptying of the vats at this time could be analogous to the Babylonian custom of cleaning granaries prior to spring harvest. Certainly the same mysterious leavening and emphasis on the coming season is inherent in both practices.

At the end of this chapter, may I indulge in a bit of poetic license? A. E. Housman said, "Malt can do more than Milton can to justify God's ways to man." Or, according to Fitzgerald, man needs only four things for paradise:
—a book of verse (beneath a bough)
—a loaf of bread,
—a jug of wine.
—And thou.

If I may take Housman's suggestion and substitute malt for Milton's poetry, let it be said that in two chapters I have already taken you three-quarters of the way to paradise.

3
CURDS AND WHEY

I am sure that when I first learned that oft-repeated doggerel about poor, frightened Miss Muffet, I had no conception of curds, whey, or, for that matter, what in the world a "tuffet" was. Vaguely, I remember an illustration in a book of Mother Goose rhymes showing a little girl sitting on an oversize mushroom about to be molested by a spider. I suppose I first thought about the relationship of curds and whey to microbiology when I began to teach. So the little girl was eating sour milk! However, I was not sufficiently intrigued then to learn why.

When I decided on the title for a chapter on cheese, I felt that my readers deserved some definitive answers on Miss Muffet's behavior. For those among you who believe that nursery rhymes are fictional nonsense, I have a surprise: Miss Muffet may have been real! I share with you the results of my diligent research. Sir Thomas Muffet was born in 1553, and although he became a physician, he did not limit himself to the practice of medicine. Before he died in 1604, he wrote several books: a book of poetry, *Silkworms and Their Fleas*; one on insects, *The Theater of Insects, or Lesser Living Creatures*; and lastly, a book of nutrition, *Health's Improvement, or Rules Comprising and Discovering the Mature Method And Manner of Preparing All Sorts of Food Used in This Nation* (England).

He was not just a dilettante in these avocations. Contemporaries considered him the "Prince of Entomologists."

More specifically, he was an expert on spiders. In this book on nutrition he gives sound advice on the eating of cheese: Don't eat old cheese, it will constipate; fresh cheese is healthiest (what could be fresher than curds in whey!). Although he married twice, he had only one child, a daughter named Patience. We have all the ingredients; the interpretation of the poem I leave to you.

The origins of cheese are undoubtedly as ancient as those of bread and wine, and indeed, references of a direct and allegorical nature are also found in the Bible: In 1st Samuel:18, Jesse tells David to "bring these ten cheeses to the captain of their thousands". And Job (Job 10:10) complains to God about his lot, saying, "You poured me in milk and poured me out cheese."

Legend has it that cheese was discovered by a Bedouin traveler who filled his kid stomach pouch with milk, mounted his camel, and set off on his day's journey. When he stopped for lunch at midday, he found that his milk was filled with lumps. In desperation, he ate the lumps and quenched his thirst with the remaining liquid. He was delighted with his gastronomic discovery. This story is now apocryphal, makes good telling, and is probably close to being a true account of what happened.

What is cheese? Most names for cheese are derived from the Latin *caseus*: Anglo-Saxon *cese*, the Spanish *queso*, German *Käse*, the Italian *cacio*. The French *fromage* and the Italian *formaggio* mean literally "formed"—relating to the fixed shape given milk solids. An excellent and complete definition is given by Davis in his book *Cheese—Basic Technology*:

> Cheese is the curd or substance formed by the coagulation of the milk of certain mammals by rennet or similar enzymes in the presence of lactic acid produced by added or adventitious micro-organisms from which part of the moisture has been removed by cutting, warming, and/or pressing, which has been shaped in moulds and then ripened by holding for some time at suitable temperatures and humidities.[1]

Ancient man found that coagulated milk, whether fresh or ripened, was an excellent way to preserve its nutritive properties from complete spoilage. Nowithstanding that, many people consider cheese as rotten milk rather than its ultimate in gastronomic expression.

As with wine and to a lesser extent with bread, cheese is a product of environment, and it is extremely difficult to reproduce a type of cheese in a locale different from its orginal source. The two major factors contributing to the final product are the nature of the milk and the ecology of cheese production. Obviously, these simply stated factors cover a multitude of controllable and noncontrollable events. The nature of the milk is the result of the strain of the animal, the season of the year, the species or kind of food, and the time of milking (morning or evening). The ecology begins at the time of milking, covers all aspects of production up to the final ripening, and includes the ambient temperature of the milkshed and such things as the resident microbial flora in the ripening rooms. However, modern technology, especially in this country, has been able to control most of these variables.

Although there are exceptions, most cheeses, like wines, are named for the region of their origin, for example, *Roquefort* in France and *Romano* in Italy. In addition, you might see the name of the animal from which the milk was derived: prefix the cheese with *Zieger* (German for "goat") or *Pecorino* (Italian for "sheep"); or a name that describes the appearance, as in *Danableu* (Danish blue); or a made-up name, *Liederkranz*.

I have made much of the variety that exists in cheeses around the world, and indeed, well over two hundred are indigenous to France alone. However, all of these can be placed in a limited number of categories, which greatly simplifies any discussion of the processes involved in their production. These categories are based on textures or firmness (hard, medium, or soft) and microbiologically whether they are ripened with mold or bacteria.

The first quality is controlled by the amount of moisture—obviously, the lower the water content, the harder the cheese. In detailing how cheese is made, I will try to make the process of moisture loss evident. Cheese begins when casein, the major protein in milk, precipitates out of solution as the milk becomes sour. This occurs as certain microorganisms, either normal to the milk or added externally (starter), grow and produce acid. To facilitate coagulation of the casein and inhibit its digestion, an enzyme called *rennet* (derived from the stomach of nursing calves) is added.[2] This produces the curd. The liquid squeezed out during the formation of the curd is called *whey*. For many varieties, especially the harder cheeses, the curd is cooked in its whey—the temperature reached not only determines final texture but also the microbial survivors, which are involved in ripening. With or without cooking, the curd is cut with special knives to release as much whey as possible, the degree of cutting affecting the final product. Lastly, the curd is sometimes pressed to remove even more whey and is placed in a suitable container for ripening. The curd can also be more finely milled and then salted before storage.

The final product is a function of time, temperature, and humidity. Although not essentially defined as "cheese," the remaining proteins in the whey (lactalbumin and lactoglobulin) can be coagulated and are traditionally used in some countries, for example, *quarkkäse* and *ricotta*.

THE EYES HAVE IT

Since it would be a practical impossibility to detail the methodology for all the world's cheeses, I am using poetic license and selecting a handful of the most distinctive, and perhaps the best known as well as representative, of the categories already mentioned. Perhaps the most famous hard cheese in the world is the one originating from the Emmen Valley of Switzerland. Allow me to digress. Several years ago, while I was in Geneva attending a scientific congress, I stopped in a little food shop for some local snacks. The first thing that I spotted was a huge square of "Swiss cheese." When my turn came to be waited on, I repressed the words "Swiss cheese." What do I call it in Switzerland? I pointed and said, "S'il vous plait, ce fromage-là avec les yeux." ("Please, that cheese with the eyes.") How embarrassing for a microbiologist not to know! Well, the Swiss call it *Emmentaler*. (See Fig. 22)

What are the origins of this cheese with the big eyes? Swiss cheeses are from big wheels with an average weight of 200 pounds. Each wheel requires about 1200 quarts of partly skimmed milk from morning and evening milking from cows that are pastured in the slopes and in the valleys.

Let us look at the microbiology of this cheese. At the beginning, the process is the same for almost all cheeses. The starter culture, containing a mixture of lactic streptococci or lactobacilli; for example, *Streptococcus cremoris* and *Streptococcus lactis* and *Lactobacillus bulgaricus*, converts lactose (milk sugar) to lactic acid as the temperature is raised to 30°C (86°F). The acidity is increased while rennet is added and the temperature is raised to 38°C. Now the casein coagulates and settles out.

At this point the coagulant (curd) is cut. Prenatal Swiss cheese is cut with a harp-like knife to particles one-eighth inch in diameter and the curd is then scalded to a temperature of 52° C (125°F). This is rather high and is chiefly responsible for the rubbery quality of the cheese. In addition, the heat selectively allows the survival of those species of bacteria responsible for the final flavor. Also, the higher the scalding temperature, the more elastic the curd. Thus the preparation of Swiss cheese contributes to the stringiness of this cheese when subsequently melted.

Two types of bacteria survive the scalding of the curd, and are responsible for the final development of flavor: they are *Lactobacillus helveticus* and several species of the genus *Propionibacterium*. The *helveticus* species (literally, the Swiss

22. Switzerland "Swiss Cheese" Shop

lactobacillus) is most active at higher temperatures (about 40°C) and produces lactic acid from lactose. The species of *Propionibacterium* have a metabolic proclivity very limited among microorganisms: the ability to oxidize lactic acid to pyruvic acid to get energy for their growth.

After scalding, the curd is collected in a coarse cloth (cheesecloth), transferred to a circular hoop, and kept overnight to allow for whey drainage. During this time pressure is applied, causing the curd particles to mat together, producing the homogeneity associated with the final product.

Allow me another digression. I mentioned earlier that genuine Swiss cheese was made in wheels of about 200 pounds weight. Although the historical reason for the size had nothing to do with the production of the cheese, bigness does contribute to the uniformity of flavor and texture in the ripened, cured product. I will explain. In medieval Switzerland, road tolls were collected from dairymen on the number of cheeses in their carts rather than the total weight. These wily Swiss increased the size of their cheese wheels, while at the same time decreasing the number of wheels. Thus they avoided the payment of higher tolls!

The ripening and curing of Swiss cheese is a fermentative process requiring no air, so that the bigger the wheel, the less this fermentation is affected by air outside. After pressing, the cheese is placed in a brine solution for several days at about 13°C (54°F), then removed from the brine and kept at the same temperature, but at a high relative humidity. During this time the surface is occasionally wiped with salt water. Susequently, the salted cheese is placed in the curing room at 20°C (68°F) where the *Propionibacteria* grow best; these bacteria produce the eyes customarily associated with the cheese. This takes six to ten weeks. What happens during this time? *Propionibacteria* ferment lactose, oxidize lactic acid, and produce a variety of end products that are said to contribute to the inimitable flavor of the cheese. Notably among them are propionic acid. In addition, other acids are produced in small amounts from the breakdown of butter fat.

At this point I should mention the importance of fodder to the final cheese. Milk from cows fed silage (stored fodder in which some microbial action has taken place) may contain an excessive number of butyric-acid-producing bacteria, the species *Clostridium butyricum,* which will not only survive the scalding temperatures but are much more resistant than *L. helveticus.* The end result is a product with much too much butyric acid and the flavor and odor of rancid butter.

Last but not least among those substances excreted by the *Propionibacteria* is carbon dioxide. Although it adds no flavor, the amount produced is a measure of the extent and quality of the metabolic reactions carried out. The carbon dioxide seeks out weak spots in the curd, pushes it aside, and forms holes or eyes usually within one inch of the surface. Obviously, the size of the holes is related to the texture of the curd mat, since the firmer the mat, the more resistant and the smaller the holes.

The metabolic pathway by which the *Propionibacteria* form propionic acid and carbon dioxide was explored by Wood and Werkman in 1941, and is a subject of broad biological significance (it will be discussed further in the Appendix). Finally, after eye formation, the wheels are cooled to 13°C (54°F) and ripened for three to six months, during which time the characteristic rind forms. The Swiss cheese is now ready for the table.

CHEDDAR IS CHEESE IN ENGLAND
In terms of volume consumed, no cheese exceeds Cheddar and its derivatives in the English-speaking world. Although the cheese gets its name from the village of Cheddar in Somersetshire, very little is made there and indeed, there is no record that great quantities were ever produced there. More likely than not, it was the marketing center, and this gave the name to that famous cheese.

There are several basic differences in the methods used for making Cheddar and Swiss. Although raw milk must be used for Swiss cheese, pasteurized milk is normally used for Cheddar varieties. Thus the normal microbial flora differ. Even though the starter may vary slightly as to species, it is a truism that the environment is much more of a factor as to what species eventually express themselves. Thus the scalding temperature for Cheddar is only 39°C (102°F), so that a different breed of survivors is involved in ripening and curing. The cutting and comminuting (reducing in size) process and equipment also differ from that of Swiss cheese, but of greater import is that the milled curd is *salted* and *then* pressed into hoops. After three to four days the formed cheese is dropped in paraffin. Cheddar cheese is either cold-cured at 0-5°C (32-41°F) or warm-cured at 10-15°C (50-59°F). The higher curing temperature usually results in cheeses with sharper, more pronounced flavor than does cold-curing.

What happens microbiologically during curing and

ripening? Fresh Cheddar cheese has a rather elastic texture with slight acidity and mild aroma. As ripening proceeds, a certain amount of the milk protein is converted by proteolytic enzymes to soluble nitrogenous compounds such as small peptides and amino acids, with a resulting drop in the firmness of the curd. In addition, a number of other products are formed that contribute to typical Cheddar flavor: for example, lactic, acetic, and butyric acids as well as smaller amounts of caproic and caprylic acids, alcohols, and esters.

The organisms primarily responsible are *Lactobacillus plantarum* and *Lactobacillus casei* which sometimes reach levels of 1 billion per gram of aged Cheddar cheese. Incidentally, the yellow-orange color of this cheese in not natural. Perhaps the dye most frequently used is *annatto*, obtained from the fruit of the annatto tree, *Bixa orellana*.

Last but not least, some mention should be made of pasteurized or processed cheese, sometimes called American cheese, which has been with us now some sixty years. I quote that renowned gastronôme, André Simon (who left us in his ninety-second year in 1970); in his book *Cheeses of the World*[3], he said that "processed cheese is a moron-like, rindless child by sterilization out of tinfoil; economical, dependable, and has many other good points; all it lacks is personality and cheese appeal." Enough said.

Swiss and Cheddar are prime examples of cheese internally ripened by bacterial activity, as are Provolone, Caciocavallo Siciliano, Pecorino Romano, Sapsago, Edam, Gouda, and Parmesan.

FOLLOW YOUR NOSE

There is another group of cheeses that are somewhat softer in texture, indicating a greater degree of proteolysis, and with perhaps a little ammoniacal pungency. With these cheeses there is a slight amount of surface bacterial activity. Into this group fall Port du Salut, Muenster, Bel Paese, Tilsiter, and Brick. When surface bacterial activity is so aggressive that a slime is produced over the outside of the cheese, and so assertive that the palates of the less adventuresome are overwhelmed, you have the cheeses of which Limburger is the archetype. These cheeses are characterized by extensive slime formation on their surfaces during ripening, resulting in a great deal of protein breakdown.

Limburger cheese originated in the city of Limburg, Belgium, and has been much copied; in fact, the German varieties of Romadur and Schlosskäse are equally infamous. This cheese is made of raw or pasteurized whole milk to which streptococcal starters are sometimes added. Rennet is used to form the curd which is heated no higher than 35°C (95°F). The curd is cut without forcibly expressing whey and placed in rectangular perforated boxes which hold one to two pounds. Salt is added to the surface, and the curd is incubated at about 20°C (68°F) and 95 percent relative humidity. The cheese is rubbed daily to distribute surface growth. After a few days, a reddish slime develops, and in approximately two weeks the cheeses are wrapped in foil and parchment. They are then transferred to 10°C (50°F) where ripening proceeds for an additional six to eight weeks.

Microbiologically, the first action is by salt-tolerant yeasts that decrease the acidity of the curd from the outside-in by digesting lactic acid. This activity permits the initiation of *Brevibacterium linens* (often called *Bacterium linens*) growth. This organism is highly proteolytic and is responsible for the reputation of Limburger cheese as among the most aromatic. Its many varieties tend to be more or less aromatic depending upon conditions of production.

One such variety deserves mention. Emil Frey, a German immigrant, tried to reproduce Schlosskäse in the town of Monroe, New York, where he lived. He did not succeed; however, what he did make is considered by some to be the best of all native American cheeses. The glee club to which he belonged thought so highly of it that they sang its praises. In honor of their melodic rhapsodies, he christened his cheese *Lieder-kranz*, literally, "wreath of song." This cheese, ready to eat about two weeks after it is made, is a delightful domestic entry on the gastronomic scene. Try some on thin sliced Pumpernickel, liberally sprinkled with chopped fresh onion and washed down with a cold glass of beer!

THE MAGIC OF THE MOLD

Now we come to the mold-ripened cheeses! Perhaps there is more mystique and legend associated with these than with any others thus far discussed. The most famous of these cheeses is Camembert. The story, by now apocryphal, has it that Napoleon Bonaparte passed through the town of Camembert in France and was treated to a singular cheese treat by one Mme. Harel. It was *the* Camembert cheese, and old Bony was ecstatic. The town became famous, the cheese became famous, and, to a lesser extent, so did Mme. Harel. In fact, a statue was erected in her honor in the town square (Fig.23)—since decapitated by military activity during World War II. However, there is strong rumor that it will be replaced despite the rather overwhelming evidence that Camembert cheese was made long before Mme. Harel and her encounter with Napoleon.

Camembert is made both in its place of origin, Denmark, and in the United States, from raw or preferably pasteurized whole cow's milk. A very active lactic-acid-producing starter is added, followed by rennet, forming a very firm curd which is not cooked but is placed in metal molds five inches in diameter and two inches deep. The molds are perforated to allow whey drainage and kept at about 20°C (68°F) and 90 percent relative humidity. As soon as possible, the surface of the cheese is salted. This helps produce a crust and also removes internal moisture. The acid content is critical, for its level prevents the growth of undesirable organisms, while the salt content and the depletion of lactose cause the die-off of the starter organisms, which may reach 1 billion/gram.

23. The Monument to Mme. Harel, the Creator of Camembert
Cheese, France
Reprinted, by permission, from Prosper Montagne, *Larousse Gas-
tronomique,* ed. Nina Froud and Charlotte Turgeon, collaborator Dr.
Gottschalk (London: Hamlyn, 1961), p. 237. Courtesy of Librairie
Larousse, Paris, France.

Shortly after salting, when the cheese cakes can be easily handled, they are dipped into an inoculum of spores of *Penicillium camemberti*. Although a number of species of yeast and bacteria can be isolated from ripening Camembert, it is the mold giving its name that is most responsible for the distinctive flavor. Growth of this fungus is evident four to five days after inoculation and reaches a maximum within twelve days. At this time there is a softening below the rind which progressively proceeds toward the center as the cheese matures. This softening is due to a digestion of casein to soluble peptides and amino acids which, at approximately twenty-eight days, is 80 percent complete. At the same time, the acidity produced by the starter is more than neutralized by the alkalinity of the amino acids.

Another French cheese similar in type to Camembert is Brie of the Marne Valley. Although not as imitated or as legendary (but perhaps more coveted by gourmets), it has earned the appellation "King of Cheeses."

If Brie is the King of Cheeses, then Roquefort is the "Cheese of Kings." Roquefort cheese is literally the reason for this book. When I first began teaching microbiology, my wife had embarked on a potential career in elementary science teaching. She presented a lesson to her classmates in science education, demonstrating how material in the home could illustrate microbial activity to youngsters. Her prime example was a sample of Roquefort cheese. "The blue veins," she said, "were a mold called *Penicillium roqueforti* growing in what was cottage cheese." She was challenged by the class, who would not believe this was the truth. "Amazing," I thought, "didn't everyone know that?" I resolved then and there to write a book educating the public on the miracles of the microbes.

Roquefort cheese got its name from the caves of Roquefort in south central France.[4] Tradition has it that some neolithic shepherd boy retreated to the coolness of one these caves to enjoy his lunch of ewe's milk curd. The bleating of his sheep suggested the possibility of a wolf in the vicinity; he abandoned his food to rescue his flock and did not return to that same cave for many weeks. He returned to find his cheese marbled with blue-green, smelling musty and rancid. Being more adventuresome than most, he tasted it. He was more than pleased. Thus Roquefort cheese was born.

Perhaps with more justification as to its origins is the story recorded by a medieval monk chronicling Charlemagne's infatuation with a blue-veined cheese which, at the time, was both inadvertent and unpredictable. The Emperor's offer of rewards for only cheeses containing blue streaks, it is said, helped to initiate and perpetuate Roquefort.

Roquefort is only one example, if the most famous, of blue-veined cheeses, but space does not permit any discussion of their specific manufacture or attributes. Let it not be said that I have deliberately ignored the virtues of Stilton, Gorgonzola, or the salty, creamy blue of Denmark. But in many respects Roquefort is distinctive; it is made from whole ewe's milk during the lambing season. All authentic produce is labeled with a sheep's head, so avoid imitations.

A very active lactic starter is used for Roquefort as in Camembert. This acidity prevents the growth of unwanted organisms. Also as in Camembert, the curd is not heated. It is however, more vigorously handled, as in Cheddar, permitting more whey to drain. About 1 percent salt is added to the drained curd and also an inoculum of *Penicillium roqueforti* spores. The culture is usually prepared on bread slices which are ground to crumbs after the mold sporulates. The inoculated curd is placed in metal hoops, 7 inches in diameter and about 3 inches high; the surface is salted, dried, and thoroughly punctured with slender needles (Figs. 24 and 25).

Now the mystery of the veins! Storage and ripening take place in limestone caves at (Fig. 26) about 10°C (50°F) at 100 percent relative humidity.[5] About ten days after storage, the starter organisms have died off, and *P. roqueforti* beings to grow

24. Inoculation of Curd with Penicillium Spores
Figures 24–26 reproduced by permission of the Société Anonyme des Caves & des Producteurs Réunis de Roquefort.

between cracks in the curd and around the puncture holes. *P. roqueforti* is aerobic—that is, it requires oxygen. Thus, the punctures allow it to breathe. (It differs greatly from *Propionibacterium* in Swiss cheese, which does not require air.) In addition, it is both salt- and carbon-dioxide-tolerant, giving it a competitive edge on any adventitious organisms. After thirty days and up to ninety days, ripening is completed. Although surface bacteria contribute some proteolysis and starter bacteria some acidity, the piquant and peppery flavor of blue-veined cheeses is attributable, for the most part, to the lipolytic (fat-digesting) activities of *P. roqueforti*.

When I was eight or nine, I read that king of adventure stories, *Treasure Island*. I still vividly remember when Jim Hawkins left the *Hispaniola*, went ashore, and found Ben Gunn, who had been marooned by pirates on the island three years earlier. The first thing the old coot asked for was "a piece of cheese." It was a long time before I learned to appreciate poor Ben's cravings. I do now.

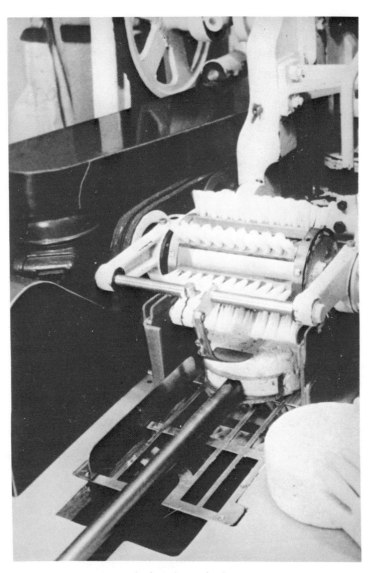

25. Puncturing and Salting the Curd Cake

26. (A) Ripening in Limestone Caves
(B) The Finished Product

4
MAN DOES NOT LIVE
BY BREAD ALONE
(OR BY WINE OR
CHEESE, EITHER)

The streak of homogeneity that prevailed in the preceding chapters is only vaguely evident here. This is the potpourri chapter—you might even say it is the leftovers. Here I will try to overwhelm you with the variety of foods consumed around the world that are subjected to some kind of microbial fermentation during their production. As you will see, a rather complete *smörgasbord* can be set with my list (Fig. 27). But however extensive it may be, I will undoubtedly be guilty of omitting someone's favorite fermented food, for which I ask forgiveness in advance.

Although there may not be the type of proof a scientist craves for hypotheses as to the origins of prehistoric customs related to fermented foods, I believe we are at liberty to make certain assumptions. Fermented foods all have three things in common: (1) after fermentation is completed, the resulting food is less vulnerable to more extensive spoilage or putrefaction; (2) the resultant products are less likely to be vectors of food-borne

A FERMENTED FEAST

Hors D'Oeuvres

Olives.....Sevillana and Kalamata

Cocktails

Scotch ● Bourbon ● Vodka
Rye ● Rum ● Gin

Wines ....of all kinds

married to an appropriate cheese
Brie..........with Champagne
Roquefort.....with Pinot Noir
Stilton.......with Port....and so on...

Sweet Cheese

Getmesost.....Feta

For Your Good Health

Yoghurt ● Surmjölk ● Kumiss
Buttermilk ● Kefir

Entrees

Oriental Delights

Fish in lao-chao (rice mash with soy sauce
Tou-fu-ru (soybean cheese)
Pidan (preserved duck eggs)...liberally with S.

Assorted Sausages

Chorizo ● Polsa ● Pepperoni
...with sauerkraut and mixed cucumber pick
accompanied by
Beer ● Ale ● Cider

Assorted Bread & Rolls

Sour Rye ● Vienna ● French
San Francisco Sourdough ● Pumpernickel

Desserts

Nata (pineapple compote, Philippine)
Brandy ● Cocoa ● Coffee

designed by chris elliott

27. Menu of a Fermented Feast

infectious disease than their progenitors; (3) the organoleptic*
changes due to fermentation have not only been accepted by
the cultures in which they were developed, but the final products have come to be treated as delicacies. As a very important bonus, these foods lose very little nutritive value in the
process, and may indeed in some cases improve.

I submit that ancient man was first taken with the
preservative qualities of fermentation and that coincidentally he
learned to appreciate the piquancy of the microbially derived
products.

*Reaction by the five senses.

COCKTAILS: FIRST COURSE

Let us appropriately begin our fermented feast with a cocktail or aperitif. I have not forgotten that alcoholic beverages stronger than naturally fermented are also derived from the simple yeast. Since essentially our concern here is with the microbiology which has done its share before the act of distillation, I cannot consider this book complete without saying something about these important cultural beverages. In general, distilled spirit drinks fall into three categories: those from fermented grain (gins and whiskies); those from fermented fruit juices (brandies); and neutral spirits with which flavors are either added or extracted (liqueurs and cordials)—the last tend to be very sweet, and lower in alcoholic content.

Allow me to serve you a portion of your favorite distilled libation; let us hope it is whiskey. Although in this country *whiskey* refers to the middle distillate of any fermented grain, aged and appropriatedly blended, to most Europeans—and especially the citizens of Great Britain—it signifies only Scotch whiskey. Indeed, the word for whiskey comes from the Gaelic *uisge-beatha,* meaning "water of life,"[1] which is undoubtedly a good indication of how the Scots felt about their national drink. Traditional Scotch was, but no longer is, 100 percent malt mash distilled. Thus the initial steps for all whiskey, including the microbiology, are the same as for beer. For Scotch barley is not only a source of necessary amylolytic enzyme, but also it serves as the sole fermentable carbohydrate. Barley is to the highlands what the Pinot Noir grape is to France. Immortalized by many bards, I am sure you all remember Roberts Burns's testimonial to the advocacy of malt mash: "Inspiring John Barleycorn, what dangers thee canst make us scorn!"

What gives Scotch its memorable quality?[2] The barley, grown amidst the highland heather; the chilly dampness of the Scottish climate; the inimitable waters from granite springs; and the dark peat moss used to fire the stills and char the aging barrels. It is this latter that gives Scotch whiskey its smoky flavor. Sad to say, traditional 100 percent barley malt whiskey is fast becoming only a legend. Today, most Scotch is blended so that other grains as well as barley are fermented. Present blended Scotch need only have a minimum of 51 percent barley malt, the rest of the carbohydrate being from other grains. In this country, the designations Bourbon and Rye have the same minimal legal restrictions—Bourbon must be made from at least 51 percent corn; Rye from at least 51 percent rye. In both cases, the malt usually does not exceed 15 percent, and may be rye or wheat malt rather than barley malt.

Whether you have chosen Scotch, Rye, or Bourbon whiskey, or the more highly distilled vodka and gin, or perhaps distilled fermented molasses, rum, you will need a piquant accompaniment for your palate.

THE OLIVE

The olive, one of the oldest cultivated fruit trees, is mentioned many times in the Bible.[3] To the ancients the olive fruit symbolized fruitfulness, and the olive wreath, peace. How strange, then, that the ripe, fresh-picked fruit has nothing to recommend it, but has an intense bitterness with no pleasant attributes to redeem it. It is probably this bitterness that led to olives being pickled, since the first step of treatment of the fruit uses a dilute alkali that chemically removes the bitter taste. It would also seem logical that the major use of the olive fruit was its oil, and that its preservation was an afterthought. Indeed, the oil of the olive was and still is the premier eating oil of the peoples of the Mediterranean basin. In fact, Spaniards in whose country 35 percent of the world's olives are produced, call them *aceitunas,* literally, the "oily ones".

Two varieties of olives, Manzanillo and Sevillano, both originating in Spain, are the major sources of green olives. These olives are harvested when full grown, but before any color changes occur. They are then soaked in 2 percent NaOH (sodium hydroxide) for several days, then washed free of alkali, and brined in about 12 percent NaCl (salt). Small casks of brined olives are set in the sun to allow fermentation to begin. The initial reactions are carried out by *Enterobacter aerogenes,* which produce excessive bubbling from CO_2 and hydrogen and acetylmethylcarbinol, and which may add to subsequent flavor. They are succeeded by mixed species which produce lactic acid from the utilizable carbohydrate. Fermentation ends when the carbohydrate is gone. At this point, lactic acid should be at least 0.5 percent and the pH no higher than 4 in order to prevent subsequent spoilage.

The Kalamata, the major black variety of olive, originated from Greece. The dark color results from NaOH treatment of olives that are beginning to ripen. At the time of plucking they are a straw to pale red in color. Aeration during alkali treatment results in an oxidation product which is black in appearance. This olive can be packed fresh, but if not preserved or pasteurized, will undergo the same fermentation as the green. These brined, oily black olives are often called "Greek olives" to distinguish them from the Spanish or "green" olives.

SWEET CHEESE

With either variety of olive, the perfect accompaniment is a fresh curd cheese (prepared without rennet), resulting just from the lactic souring of milk. Orthodox Jews must restrict themselves to these fresh cheeses because dietary laws prevent the mixing of milk with products derived from meat (rennet comes from calf stomach). Cottage cheese, pot cheese, farmer cheese, clabber cheese are all examples, but the companion par excellence from Greece is Feta, goat's milk curd, preserved under brine. Or perhaps you should try a rather different cheese product, though not strictly a cheese: Getmesost, made of cooked whey solids. It has the color of butterscotch, the overwhelming pungency of burnt milk, and is considered a Swedish delicacy. It actually tastes sweet!

AND LIQUID REFRESHMENT TOO!

And for your beverage, why not some fermented milk? Purposely soured milk as a beverage is drunk throughout the world, and indeed has become a part of many cultures. Basically two types of fermentation are involved: (1) bacterial lactic acid and (2) mixed bacterial lactic acid and yeast ethyl alcohol. The variety of milk-yielding species stretches from the Ass to the Zebra, but perhaps the most notorious of fermented milks comes from the Balkans and most probably from nanny goats. This notoriety can be traced to the health fad craze which started after World War I. Elie Metchnikoff, who had shared the Nobel Prize in medicine for his pioneer work in immunology, became obssessed in his later years with the reasons for the apparent longevity of Bulgarian peasants. He ultimately attributed this longevity to the continual drinking of yogurt, a type of local fermented milk. He suggested that the lactic-acid bacteria in the soured milk replaced the normal intestinal organisms whose metabolic activities were partly responsible for tissue breakdown that led to aging.

Although neither his premise on longevity nor the reasons for it were ever scientificallly demonstrated, his suggestions resulted in a spate of products on the world scene. Millions of people imbibed fermented milk as readily as if it came from Ponce de Leon's Fountain of Youth. I still remember the small green bottle (similar to a 6 ounce soda pop bottle) of Fermalac (a brand of so-called acidophilus milk) left at our doorstep daily by the milkman with his regular order of milk. Although the intestinal flora are never permanently displaced, sour-milk drinking has found a place in the national diet.

The fact that, microbiologically, two broad types of sour milk exist, is an accidental function of the source of the milk, the ambient temperature of the area where it was made, and the starter's usually being the dregs of the last batch. So from Russia we have Kumiss (a drink derived from mare's milk containing 2 percent ethyl alcohol and 1 percent lactic acid) and from Sweden Surmjölk (a thickened lactic-acid drink, possibly from reindeer milk).

However, there are three kinds of soured milks based not on the *origin* of the milk, but on its chemical nature, and are all produced in the United States. Perhaps the most popular is buttermilk. Buttermilk is literally the fluid remaining after the cream is churned into butter. The beverage consumed is cultured buttermilk, prepared by souring true buttermilk. Today it is common in commercial dairies to add butter starter to skim milk; quite often, a small amount (1 to 2 percent) of butter flake is added to enhance the flavor.

Two groups of bacteria are involved in buttermilk fermentation: *Streptococcus* and *Leuconostoc*. The former are primarily responsible for producing lactic acids (and the resulting sour taste), while the latter produce the flavor and aromas once the acidifiers have prepared the way. In fact, the *streptococci* produce some important growth factors without which some *Leuconostoc* species cannot grow.

In any event, the metabolic products that give buttermilk its unique flavor result from the further breakdown of citric acid—these are acetylmethylcarbinol, diacetyl and acetic acid —and are produced after a starter inoculation of the pasteurized precursor and incubation at 69°F until a level of 0.8 to 0.9 percent acid is reached.

Acidophilus milk, the other sour-milk product that is consumed, although to a more limited degree than buttermilk, is derived from inoculation of sterilized skim milk with *Lactobacillus acidophilus* cultures. This is a normal member of the intestinal flora of man, especially infants, and is thought by some physicians to be a cure for certain kinds of gastro-intestinal upsets. Some say this species establishes itself as the dominant intestinal microbe, bringing order out of chaos. These presumptions might appear to give credence to the original proposal of Metchnikoff. However, the use of acidophilus milk or even pure tablets of *L. acidophilus* cultures is not sufficiently widespread to make any decisive claims for it.

Yogurt, the third type of sour milk, is experiencing a revival in consumption. Yogurt differs from the previous two products in that it was originally prepared from whole milk boiled down to one half its initial volume. However, a similar effect (more or less) can be achieved by adding 4 to 5 percent dry-milk solids to whole milk, or by using condensed milk. Another distinction is the heat treatment, which can be as high as 200°F for as long as 90 minutes. This not only lowers the resident microbial flora level but also makes the milk more suitable for growth of the yogurt culture. After cooling, the yogurt is inoculated with 2 to 3 percent starter, placed in small dispensing containers, and then incubated at 113°F. The rather large inoculum produces the desired acidity (about .9 percent) in about 3 hours.

Two species of bacteria are responsible for yogurt: *Streptococcus thermophilus* and *Lactobacillus bulgaricus*. There is no general agreement on their relationships other than both species appear necessary to produce both acid and flavor. Incidentally, yogurt was originally made from goat's milk, but today it is made from cow's milk; because of the current preocupation with calories and cholesterol, skim or partly skim milk is used in its production.

ORIENTAL DELIGHTS

Let's change directions somewhat, and go East. The substrate for most oriental foods is either soybean, rice, wheat bran, or a mixture of all three. The major microbiological distinction in their fermentations is the involvement of at least one species of mold. The starter, termed *koji* by the Japanese and *chou* by the Chinese, requires a mixture of appropriate microbial species with starch-, protein-, and fat-digesting enzymes (see Fig. 28). Although bacteria and yeast are also involved, the important species is *Aspergillus oryzae*. A moistened mixture of soybeans,

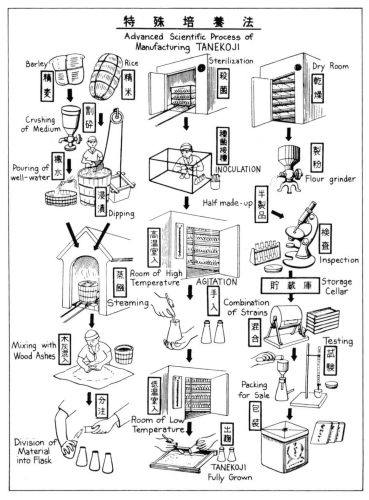

28. *Koji* Starter
An adapted reproduction from *Tanekoji,* courtesy of Kojiya Sanzal-mon Roho Ltd., Kyoto, Japan.

rice, and wheat bran is inoculated with *Aspergillus oryzae* and incubated at about 86°F. During this time lactic-acid bacteria, as well as proteolytic and amylolytic species of *Bacillus*, develop, e.g., *Bacillus subtilis.* This starter is used in a variety of products. The list of possible offerings includes condiments and beverages, as well as main courses from China, Japan, Indonesia, and India. I must, therefore, be selective.

Perhaps, the foremost product is "soy sauce," the condiment par excellence of the Orient. It is made from hydrolyzed soybeans mixed with roasted wheat and steamed wheat bran. This mash is inoculated with the koji and incubated at 86°F in open trays for three days, soaked with 24 percent NaCl solution, and held as long as one year. The aging and fermentation is a complex affair. The lactic-acid bacteria produce sufficient acid to prevent spoilage; species of yeast produce small amounts of alcohol which help determine the flavor; *Bacillus* species decrease turbidity by digesting proteins and starches; and last but not least, *Aspergillus oryzae* contributes the predominant flavor and aroma. For the microbiological mysteries of *tamari* and *miso* and of *ragi-roti, ragi-tempe,* and *ragi-ketjap,* please consult the references in the back of this book.

ENTREES

As the main course you may choose *tou-fu-ru* (Chinese soybean cheese), a bean curd fermented with a mold of the genus *Mucor,* or fermented fish strips preserved in *lao-chao* (fermented rice mash), or *pidan* (preserved duck eggs), fermented during storage by adventitious *Bacillus* species and so-called coliform organisms. Accompaniments include *poi,* a mixed fermentation of the corms of the taro plant involving heterofermentative lactic-acid bacteria, *Pseudomonas* species, yeasts, and a mold—*Geotrichum candidum.* Wash this all down with *sake,* that delicately bouqueted rice "wine" made by hydrolyzing rice starch with *Aspergillus oryzae,* then fermenting the sugars to ethyl alcohol via *Saccharomyces* yeast.

Turning your attention from the Orient to Europe, you encounter a variety of sausages: Pepperoni from Italy, Chorizo from Spain, Thuringer from Germany, Pölsa from Sweden—all cured with a mixed lactic-acid fermentation that not only acts as a preservative, but also contributes to final flavor and aroma.

What better accompaniment for these tangy meats than sauerkraut and pickles! Sauerkraut is unique among pickled foods in that it is usually eaten as a main-course vegetable rather than just an accompanying delicacy, as with other pickles. The origins of pickled cabbage are obscure, but they undoubtedly derive from central Europe where it is still a mainstay of Polish and German diets. In fact the French, who rarely defer to anyone, particularly in things culinary, call sauerkraut (German for "sour cabbage") chouxcroute (literally "cabbage cabbage").

Sauerkraut is not only a well-preserved form of cabbage and a tasty addition to the table, but also a nutritional tour de force. The Vitamin C content of sauerkraut is equal to that of citrus fruits and, in fact, was taken along on the voyages of Captain Thomas Cook in the 1770's to prevent scurvy.

Manufacture of sauerkraut is simple and inexpensive. Shredded cabbage about one-sixteenth of an inch across is mixed in an appropriate fermentation vessel with 2 1/2 percent NaCl, based on the weight of the cabbage. The amount and incorporation of NaCl is extremely critical. The high salt concentration ruptures the cells of the cabbage leaf, releasing the tissue juices which are subsequently fermentable. Important procedural aspects are both the covering of the brined cabbage in order to exclude as much air as possible, and suitably sinking the fermenting mass until all of it is below the liquid surface. A day or so after brining, fermentation begins with cabbage microbes metabolizing the nutrients extracted from the leaves. The fermentation is temperature-dependent (45° to 70°F) and takes from three to six weeks, eventually reaching about 2 percent acid. It is very important that air be excluded from the finished product, since certain molds and yeasts can degrade the acids formed, causing spoilage.

Three acid-forming species of bacteria are mainly responsible for carrying out normal sauerkraut fermentation.

The first species is *Leuconostoc mestenteroides,* a *coccus* which produces acetic and lactic acids, ethyl alcohol, and large amounts of CO_2. This latter is very important, since it creates the anaerobic environment necessary for optimal fermentation. Subsequently this species is killed by its own acid production and is succeeded by *Lactobacillus plantarum,* which produces only lactic acid. It, too, eventually commits environmental suicide. The last species, *Lactobacillus brevis,* is extremely acid-tolerant and completes the fermentation. The final product contains lactic and acetic acids in the ratio of four to one.

In our society, *pickle* is practically synonymous with *pickled cucumber.* Some thirty-seven types of dill, sour, and sweet pickles have been catagorized (see Appendix), but essentially the fermentations are all the same—only the auxiliary herbs or post-pickling procedure being different. It is important that good-quality cucumbers be used and that the pickling water not be very hard. Brining is extremely critical: The NaCl level should not be below 8 percent, preferably about 10.5 percent. In production, a brining solution equivalent to 9 lbs. of salt for every 100 lbs. of cucumbers is used. As cucumber-tissue water is leached by osmosis the brine becomes diluted; adequate salt levels must be maintained to prevent spoilage. The brine produces a color change in the cucumber from bright green to olive, and is responsible for preserving the internal integrity of the fruit.

The curing or pickling is the result of the metabolic activity of three groups of microorganisms: *Aerobacter aerogenes,* the lactic-acid bacteria, and some terminal yeasts. During the first twenty-four hours *Aerobacter aerogenes* dominates and produces large amounts of CO_2 and hydrogen. This establishes anaerobic conditions and paves the way for the lactic-acid bacteria. After one week *A. aerogenes* is gone and *Lactobacillus plantarum* is in ascendancy. For about three weeks it forms an abundance of lactic acid. At this time some yeasts of the genus *Torulopsis* appear and may be responsible for certain flavors associated with ripe pickles. Pickles can take up to six weeks to cure depending on the degree of sourness desired. In addition, other herbs and spices such as dill and garlic can be included for variety. However, these are only auxiliary and have no effect on the fermentation. The finished pickle is a fragile food subject to deterioration by a number of microorganisms capable of utilizing lactic acid. One sour dill pickle referred to as "Kosher style" traditionally is consumed au naturel —that is, without any further treatment to prevent spoilage, although pickles are usually heat-pasteurized or further acidified with 5 percent acetic acid (vinegar) to assure their keeping quality.

Acetic acid, or vinegar, is itself one of the most important products of microbial activity when desirable (when not as a result of wine souring).[4] The basis for vinegar production is ethyl alcohol derived from one of several fermentations, usually dependent upon which carbohydrate source is most readily available: for example, malt vinegar in Great Britain, wine vinegar in France, and cider vinegar in the United States. In addition, it is possible to use neutral grain spirits from which almost all traces of their antecedents have been distilled. However, it has not yet been fully proven that some trace metabolites may not indeed be necessary for the optimal growth of the acetic-acid-producing bacteria.

The microbiology of vinegar fermentation is rather straightforward. It involves the oxidation of ethyl alcohol to acetic acid by members of the genus *Acetobacter*. These organisms are the major components of the film slime known as "mother of vinegar." Actually the factor that limits the rate of this metabolic reaction is oxygen, and during commercial production of vinegar, wooden chips (birch, beech) are used to increase the surface area exposed to oxygen and also to give a holdfast for the growth of the *Acetobacter* films. Vinegar is defined by the U. S. Food and Drug Administration as 4 grams of acetic acid in 100 milliliters of solution.

If I have spoken disdainfully of fermented juices other than those of the grape, now is the time for me to retract my claws. Cider, the fermented juice of the apple, and perry, the fermented juice of the pear, have a nobility in their own right, and there is something to be said for the simple virtues of the many natural fermentations carried out in the home. Even the lowly dandelion has its advocates, while mead, from fermented honey, is no less appreciated today than when quaffed on Olympian heights by Zeus and his cohorts.

The last alcoholic drink I will mention is pulque, the fermented heart juice of the agave or century plant. This fermentation is unique. While all others form ethyl alcohol with strains of *Saccharomyces cerevisiae,* pulque is the result of *Zymomonas lindneri* metabolism (see Appendix).

And now the end of the meal! Yes, even coffee and chocolate owe their final flavors to microbial activity. Coffee is made from the berry of *Coffea arabica,* the bean or seed being originally surrounded by a fleshy pulp. The fermentation of this pulp contributes to the final aroma of the drink. One method involves the initial removal of the outer skin of the berry by pectin-digesting microorganisms (bacteria and fungi) followed by a fermentation involving our familiar lactic-acid bacteria, for example, *Lactobacillus plantarum, Leuconostoc mestenteroides,* and *Streptococcus fecalis.* For dessert, chocolate—the bean of *Theobroma* ("food of the gods") *cacao.* The slimy, fruity pulp covering the bean is removed by a sequential fermentation involving yeasts and bacteria. These reactions are responsible not only for digesting away the pulp, but for killing the seed embryo and partially determining the flavor, color, and aroma of chocolate. If chocolate is not your cup of tea, then perhaps you might

29. Copper Still Originally Used by Charles Martell in the Early 18th Century

sweeten you palate with *nata*. This Philippine delicacy, a kind of cellulosic compote of pineapple whose preparation was long a mystery, is now know to result primarily from the metabolic activities of *Acetobacter xylinum*.

Last but not least, pour yourself a ration of brandy (from the Dutch *brantewijn* meaning "burnt wine", that is, distilled wine,) (Fig. 29), into that most particular of containers, the brandy snifter. More specifically, make it Cognac, the archetype of brandies, distilled only from the white wine of white grapes grown in the valley of the river Charente, whose soul, the city of Cognac, lies 312 miles southwest of Paris. Only this distillate, aged and blended, deserves the appellation Cognac.

Select from among VSO (Very Superior Old, 12 to 17 years old), VSOP (Very Superior Old Pale, 18 to 25 years old) or VVSOP (Very Very Superior Old Pale, 25 to 40 years old); cradle your snifter, caress it, swirl it, tilt it, gently inhale the aromas brought forth by the heat of your hands, fondle the precious liquid with your tongue and give thanks for fermentation.

Bon appetit!

5
FOOD FOR THOUGHT, OR, A LINK IN THE FOOD CHAIN

A major problem associated with the ever increasing world population is that the numbers of people may one day exceed the available food supply. According to many experts, however, and the Food and Agricultural Organization of the World Health Organization, even in 1970 food production kept pace with the population explosion. This was possible through improvements in agricultural technology and innovations in plant genetics. The greatest problem lies not in the production of food, but in its distribution. Food in this context essentially means protein. Protein-poor areas of the world continue to be so. In addition, it may not be possible in the future to keep pace with population growth, and some new approaches might be necessary. Food from microbes, especially protein, offers a potential solution.

Indeed, at the Third International Fermentation Symposium held in 1968 Carl-Goran Heden, the eminent Swedish microbiologist, entitled his address: "Ferment or Perish: Future Role of Applied Microbiology in World Affairs." His intent was to bring home emphatically the case for a number of microbial activities that might aid in solving several problems of our times.

Notable among these activities is the direct use of the microbe itself as a source of single-cell protein (SCP).[1] In the first four chapters, I have discussed some of the products made by microbes and consumed by man. In this chapter I will treat a relatively new technology—the production of a variety of microbial cells for their protein content. The production and acceptability of SCP presents difficulties, however, since there are innumerable nutritional, toxicological, social, and economic hurdles as well as the technological ones.

Let us look at the potential usages of SCP. Adding it directly to man's diet has the lowest degree of acceptability. I will explain why later. However, man's rejection of microbial protein is not a significant drawback since the greatest promise for SCP is a nutritional supplement to protein-poor foods or as animal fodder, especially where grazing lands are restricted. In addition, a most novel use involves waste conversion to potential food using microbes in closed ecological systems such as spaceships and space stations. In these situations it becomes inevitable that during long-term absences from the earth's environment the recycling of nutrients and waste disposal will be one and the same operation. Research into such closed systems has been intrinsically associated with the development of the space program in this country since the 1950's.

Because man has already been eating microbes to a limited degree in the bread, the wine, and the cheeses he consumes, the total concept is not so very exotic. In the last fifty years several varieties of microbial types have been investigated as possible sources of foods; these include several species each of bacteria, yeasts, molds, algae, and protozoa. What are the advantages of microbes as food? Actually, they are no better food sources than are animals or plants. Their superiority is in how and on what they grow. The potential for microbial growth is enormous: maximal generation times for certain species are as short as ten minutes. This means a doubling of biomass within that time.

Of greater significance, however, is the relative yield of protein from a microbial system when compared to more conventional food sources. In the following table you can readily see the tremendous difference between animal- and plant-protein yields on the one hand, and those of yeast and bacteria on the other.

TABLE 2
Yields of Protein from Microbial Systems compared to Conventional Food Sources

Source	Protein Produced Per Day
1,000 lb. steer	1 lb.
1,000 lb. soybean	100 lb.
1,000 lb. yeast	100,000 lb.
1,000 lb. bacteria	100,000,000,000,000 lb.

It is their rapid growth that must be exploited. No less valid are their substrates, or what the bacteria feed on. Many species are literally capable of turning sows' ears into silk purses, having been grown on such assorted energy sources as petroleum, waste water, rancid oil-based grinding fluids, molasses, sulfite waste liquor, wood hydrolysates, and kerosene. This rather broad list suggests only what can be done in any one geographical location. For example, at Louisiana State University engineers have put the bacterium *Cellulomonas* to work eating bagasse (sugar-cane residue). When dried, the organisms contain 50 percent protein. Five pounds of sugar-cane residue yield about two pounds of cellulomonas cells, and nutritional tests indicate it has high food value. Man has known of the nutritive value of microbes for some time and, especially considering the vitamin content of yeast cells, of their therapeutic value also.

There appears to be some evidence that in the fifth century, B.C., Hippocrates recognized the curative powers of yeast cells. However, food as an energy source is quite different from its vitamin usage, and it was not until World War I that an interest in microbes as food developed. Due to food shortages in Germany at that time, the yeast *Torula utilis* (now *Candida utilis*) was grown as a nutritional supplement. Unfortunately, as with many other positive wartime exigencies, interest in SCP waned. Although yeast as a food supplement has continued to be used since World War I, no big breakthroughs in complete acceptance have occurred. Perhaps the most exciting and intriguing use of SCP has developed since 1963, involving microbial growth on petroleum fractions and pure hydrocarbons.

I mentioned earlier that even though food production has kept pace with population growth, poor distribution has kept food from the hungry. The dramatic finding that microbes could grow in petroleum was all the more meaningful when you consider that major oil deposits are located in many areas where protein deficiencies exist. I first became aware of food from petroleum in early 1964 when I was visited by a young scientist from L'Institut Francais du Pétrole. He told me about pilot operations in Lavera, France, supported by a division of British Petroleum Company, Ltd., in which a strain of *Candida lipolytica* grew abundantly on a crude petroleum fraction, literally dewaxing it (cracking it by removing the higher boiling portion), and leaving the more valuable gas oil. This indeed was having one's cake and eating it too!

My French colleague facetiously suggested that the process had more geopolitical than nutritional implications, since the willingness to build petroleum fermentors in the Arabian Peninsula might gain friendships for France even if hungry people never swallowed a single yeast cell. I was almost convinced of the truth of his tongue-in-cheek political comment some months later when, on a visit to his laboratory in France, I was politely told that it would be impossible for me to see any of the ongoing fermentation research because of its highly restricted nature. Since that time research on SCP from petroleum has spread round the world.

It was orginally considered that SCP from petroleum was ideally suited for so-called emerging nations, primarily as a human food supplement and, indeed, very active research has been conducted on plant levels in India similar to those initiated at Lavera. The petroleum is available, the technology is adequate, all that remains is to convince the people to eat it once produced. During the past seven years studies on SCP yields from hydrocarbons using not a yeast, but a suitable strain of a bacterium, *Pseudomonas* number 5401, have been carried out in Taiwan with much success. The growth rate and protein yields were greater than with yeasts, contained all essential amino acids, were fermented at a temperature suitable for a tropical ambience, and were tasteless and odorless and, therefore, more readily acceptable as human food.

Initial studies on the nutritive value of the *Pseudomonas* 5401 protein from petroleum substrates based on feedings to mice, chickens, and hogs indicate that it is the equal of soy bean protein with respect to growth, development, and fertility.

Japanese studies with the yeast *Candida tropicalis* grown on a pure hydrocarbon, n-hexadecane supplemented with ammonium nitrate, are also encouraging regarding yields of protein with the right balance of amino acids. In addition to good protein, these cells also are rich sources of the B-vitamins needed in animal and human nutrition. According to Marvin Johnson, a University of Wisconsin biochemist, it is technologically possible with known processes to utilize between 15 and 20 percent of the world's petroleum production to produce 100 percent of the protein required by the world's inhabitants. Unfortunately, this would not be acceptable economically or socially, but it does point up the potential for petroleum as food or fodder for the future.

Although many species of bacteria yeast and filamentous fungi have been successfully grown on some form of hydrocarbon derived from petroleum, there are cogent reasons for selecting certain microbial types over others. The fungi in this case should probably be avoided. (In some other instances they may be more advantageous; I will mention them later in the chapter.) Not only are they not nutritional equivalents to yeasts and bacteria (see Table 3) but the harvesting of fungi for eventual disposition imposes additional technical problems due to their filamentous nature. Between yeasts and bacteria there is little nutritional difference. However, bacteria are more difficult to separate from spent culture media because of their smaller size, and their high nucleic acid content may be toxic. What are the advantages of using bacteria to obtain SCP? They grow much faster and they have broader appetites for a variety of molecular types of hydrocarbons. This omnivorous behavior would be selective if the kind of energy source were highly limited.

How close are we to exploiting this new food resource? Are the affluent societies prepared to accept "hi-test, no-knock" canapes? Are regions plagued by kwashiorkor (protein deficiency disease from consuming only plant protein low in lysine) ready to accept handouts of petroleum protein? I will postpone the general subject of direct acceptability until the end of the chapter, once I have discussed all aspects of SCP. At the present time the major use for petroleum-derived SCP is not for direct consumption by the nutritionally impoverished in the emerging countries of the world but, according to Ivan Malek of Czechoslovakia, as a remedy to animal-fodder protein deficiency in technically advanced countries with shrinking space resources for traditional feeding practices. With this in mind, British Petroleum after ten years of planning, scheduled in mid-1971 the opening of two plants for this purpose, one in Lavera, France, and the other in Grangemouth, Scotland. They originally anticipated producing 20 million metric tons of SCP

TABLE 3
Essential Amino Acid Contents of Microorganisms (Plant and Animal Samples Compared)

Sample	Content (%)	
	Essential amino acids (dry wt)	Nitrogen
Bacteria		
Staphylococcus aureus (13)	21.6	10.75
Escherichia coli (13)	33.1	13.19
Bacillus subtilis (13)	23.8	10.07
Yeasts		
Saccharomyces cerevisiae, av. (14)	17.1-23.8	5.9-8.2
Saccharomyces cerevisiae (13)	23.1	8.94
Torula yeast (15)	29.5	8.35
Torula yeast (16)	24.4	7.47
Molds		
Aspergillus niger (13)	9.2	5.21
Penicillium notatum (13)	12.8	6.13
Rhizopus nigricans (13)	9.6	5.80
Mushrooms		
Tricholoma nudum (17)	20.8	8.64
Nonmicrobial samples		
Animal muscle, av. (18)	48.1	15.4
Fish meal, av. (16)	32.1	9.8
Alfalfa meal (16)	6.9	2.72

*SOURCE: Marvin J. Johnson, "Growth of Microbial Cells on Hydrocarbons," *Science* 155 (March 24, 1967): 1516.

from petroleum for the Western European animal feed market at a price competitive with soy bean and fish meal protein. However, changes in the world petroleum market have made exact production figures difficult to come by with regard to their operations. The oil-producing countries have raised the price of crude petroleum so high that price advantages from the production of single-cell protein may now be lost. It is a sad commentary that so many countries with food needs that could possibly have been solved by this new technology may now be caught in a political squeeze. A continental competitor of British Petroleum, Shell, is instead looking at bacterial fermentation of methane by *Methylococcus capsulatum,* which would enable them to take advantage of the huge natural gas deposits in the North Sea.

One more aspect of SCP growth deserves discussion. One of my colleagues, Professor E.O. Bennett, of the University of Houston, has literally "killed two birds with one stone" by suggesting a novel approach to a serious waste problem and converting it into a potentially useful product. For many years, the disposal of millions of gallons of spoiled oil-in-water cutting-fluid emulsion has plagued the machine tool industry.[2] The primary culprits involved in the cutting fluid spoilage are members of the bacterial genus *Pseudomonas.* Bennett reasoned that since this group was purposefully being used to produce SCP on similar hydrocarbon substrates, one could take spoiled oil emulsion, stimulate more growth, harvest the bacterial cells, and then check their nutritive value. Stimulation of growth required an added source of nitrogen. Bennett ingeniously found a cheap and readily available source; early morning male urine (conveniently supplied by his graduate students) at a level of 0.5 percent produced optimal growth. The amino acid content of the protein from one such culture had a nutritional index between chicken and beef muscle. Thus, Bennett seems to have succeeded in making one silk purse (SCP) from two sow's ears (waste oil and urine). What can be done with this potentially valuable protein? He offers some valuable advice and cites Metropolitan Detroit, Michigan as an example: this area is a major user of cutting-fluid emulsion and in addition is surrounded by a large body of fresh water. He suggests a central fermentation plant supplied by waste oil from nearby plants. The SCP yields would be used in fish farming in the neighboring lakes. Pilot studies carried out by Bennett for small ponds have successfully demonstrated that catfish can be raised with spoiled-oil SCP.

Another novel process for waste utilization developed by M. Tveit in Sweden, is worth mentioning. It is called the "Symba-Yeast Process" and refers to the use of two species of yeast together (symbiotically), *Candida utilis* and *Endomycopsis fibuliger*. The latter digests starches yielding sugars utilized by *Candida utilis*. The mixed product contains 40 to 50 percent protein, of which 80 to 90 percent is digestible. This microbial mix can be used to remove the starch from dehydrated potato plant effluent, for example. Tveit estimated that the cost of water purification by the Symba process would be about 50 cents per ton compared to six to ten dollars per ton by conventional methods.

Crude petroleum and petroleum-derived hydrocarbons are impressive examples of large-scale available substrates for growing SCP, but they are by no means the only such sources. Ideally, energy-yielding substrates should be surplus, cheap, and located where the need for protein exists. This may even be extended to using waste liquors still containing food that may be used by certain yeast strains.

Granted that direct feeding of SCP to man is not around the corner, I feel it is worthwhile to remind my reader that torula (a kind of yeast), baker's, and brewer's yeasts either intact or hydrolyzed (leached), have been used for some time as protein and vitamin supplements in the human diet. Torula fed at the rate of 130 grams/day to human subjects was assimilated at nearly the same rate as animal protein. This is a rather high dose, not yet universally accepted as nontoxic. However, an overwhelming number of reports would support the premise that yeasts are safe for consumption. The three forms of commercial yeast differ in their methods of preparation. Brewer's and baker's yeasts are both types of *Saccharomyces cerevisiae* (see Chapter II) and differ only in the processes in which they are used. Nevertheless, the processes themselves make brewer's yeast much less desirable. It is lower in B-vitamin content and must be de-bittered (to remove all traces of hop flavor) before use. Baker's yeast, sometimes called primary yeast, is usually grown on a molasses substrate. Torula derived from *Candida utilis* is preferred since it produces a greater yield of protein and vitamins for a given substrate. In addition, it has a much broader carbohydrate appetite than *Saccharomyces* species enabling growth on such substrates as sulfite liquor (waste from paper pulp manufacture). Liquors from softwood contain up to 2 percent sugar, acid, and nonsugar carbon, increasing the cell yield 10 percent above that expected on the basis of sugar content alone. The quality of food-grade sulfite yeast is specified by the National Formulary XII and by the U.S. Food and Drug Administration to contain 55 percent protein. About 1,000 tons of this food additive are produced each year in the United States. A most important use of yeast as a food supplement is as an adjunct to grains deficient in the amino acids lysine and tryptophan. It is the low level of these two amino acids in maize protein that is primarily responsible for the infamous disease of malnutrition, kwashiorkor.

SUNSHINE STEAKS

So far I have said nothing about the most obvious and what would seem to be the cheapest source of SCP—microbial cells, that is, deriving their energy from light. These are representatives of simple plants, the algae. They are primary producers of food energy, converting sunlight into chemical energy by fixing or taking up CO_2 from the environment and turning it into cell protoplasm and most importantly, protein. It has been estimated that primary productivity in the world's oceans is 550×10^8 tons of biomass annually, more than 100 tons per year for each individual in the world. Man consumes this indirectly by consuming fish and other sea animals; for example, the world's

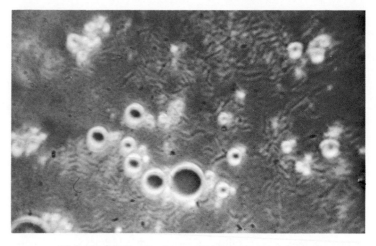

30. Spoiled Cutting Emulsion—Oil Microdroplets in Suspension
Notice the varying sized globules. These are oil microdroplets in
suspension or emulsified. When deterioration sets in, small droplets
coalesce until the oil is seen to grossly separate out to form a separate
layer. In the background, you can see the numerous rod-shaped
bacteria responsible for the spoilage.

oceans produce an estimated 48 million tons of fish annually,
all originally derived from algal energy. Would it not be simple
to harvest the oceans with huge vacuum cleaners to make use
of this giant free protein factory? Unfortunately the concentration of marine microalgae is so low, not greater than 3 milli-
grams per liter, that the cost of harvesting would far exceed the
value of the crop (the break-even point is no less than 250
mg./liter). Thus, for all practical purposes we must leave reaping
of the seas to the less cost-conscious fauna, and rely on better-
controlled artificial systems.

A number of species of green and blue-green algae
have been used in open and closed systems in small pilot and
larger production units for experimental purposes and for active
utilization. The human consumption of algae cells is not a recent development, albeit its large scale exploitation is. Re-
searchers at the Institut Français du Pétrole found that several

tribes in the African republic of Chad had been harvesting and consuming a species of blue-green algae since ancient times. This proved to be *Spirulina maxima,* which the French began investigating in 1962 for large-scale growth studies. It looks like a good SCP candidate. It contains 60 to 68 percent protein by dried weight and is one of the few organisms that grows best in fairly alkaline mediums. Natural growth is in shallow ponds containing high levels of bicarbonate.

As growth proceeds the following reaction takes place:

$$H\ CO_3^- \longrightarrow CO_2 \quad + \quad OH^-$$

(bicarbonate)   (carbon   (hydroxyl ion
dioxide)   alkalinity)

$$H_2O \quad \Big\downarrow \quad Sunlight$$

$$[CH_2O] + O_2$$

(carbohydrate)   (oxygen)

As carbon dioxide is used, the hydroxyl ion is produced making the pond more and more alkaline. Although the algae are extremely resistant to alkaline conditions, artificial ponds do not have an unlimited supply of bicarbonate to supply CO_2 for growth. The French solved this technical problem by bubbling in CO_2 from combustion. This served the dual purpose of supplying the carbon for photosynthetic growth and doing away with the need for mechanical mixing, thus reducing the cost of the process. The initial daily yields of 80 pounds of dried algae per acre containing 65 percent protein makes this a highly promising SCP candidate. Currently, the French group is comparing yields from pure cultures of *Spirulina maxima* with those from more uncontrolled open cultures which obviously are easier and less expensive to maintain.

At the Tenth International Congress of Microbiology in Mexico City, August 1970, it was reported that the use of initial pure cultures of *Spirulina* in a medium of relatively high salt content permitted open cultivation without the undue problem of contamination. One of the bonuses associated with the *Spirulina* study was also presented at that Congress. It was found to have antimicrobial activity.

Perhaps more widespread use has been made of large-scale growth of species of *Chlorella,* a green microalgae. The Soviets report that systematic feeding of 50 to 100 grams of dry algal mass per day in human diet had no deleterious effects, and in Japan *Chlorella* is being sold as an additive to yogurt, ice cream, and related products. However, there are problems with algal food. Odor and taste are inhibiting to direct consumption; in other words, they don't taste very good. A more severe drawback is the indigestibility of the cellulose cell wall, making them useless unless predigested to release the SCP. Obviously this does not restrict its use as auxiliary animal feed, particularly for the ruminants.

In this regard, the pioneer efforts of a group directed by Professor William J. Oswald at the University of California must be mentioned. Between 1960 and 1966, algae grown on waste water ponds was harvested and fed to swine, sheep, and cattle as 10 percent of their total intake, with excellent results. The predominant species at harvest was *Scenedesmus quadricauda,* which apparently has a better survival capability than *Chlorella* on waste water.

These studies at the University of California and other places point up one of the most promising and fruitful areas of SCP research: nutrient and oxygen regeneration in closed systems. The closed systems of most immediate interest are the extraterrestrial environments associated with spaceships in-

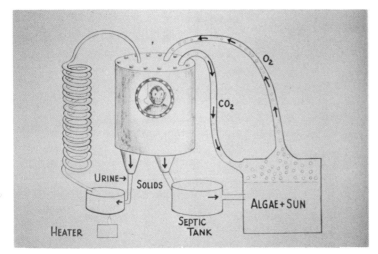

31. Scheme for Recycling of Wastes in a Closed System

tended for long trips where weight considerations prevent the transport of sufficient oxygen or food and space stations and planetary colonies where the same problems prevail. Also, we should not overlook the fact that our own planet is becoming more and more of a closed system, increasingly dependent on regeneration. Research is being directed toward mixed systems of bacteria, fungi, and algae in which wastes are partially or totally mineralized by fungi and bacteria that produce CO_2 used by algae in the presence of sunlight (which evolve oxygen etc., etc.) (see Fig. 31). Such *biocenoses* are difficult to stabilize at the so-called pilot level since individual growth rates in mixtures are relatively unpredictable. However, a group of scientists from the Siberian Academy of Sciences recently reported success with a heterogeneous microbial population.

If we superimpose man or other primates upon this community, we further complicate it. Nevertheless, such studies have been going on for some time. Dr. Robert Tischer and his group at Mississippi State University have looked at human feces as the basis for supporting algae growth as well as the pyrolysis gases produced when feces was burned. Both his results showed it was possible to grow algae with feces or its breakdown products. Growth alone, however, is not sufficient for closed systems. The optimal species should have a number of minimal qualities:

(1) High growth rate;
(2) Growth at relatively high temperatures;
(3) Growth at relatively high pH;
(4) Good food value and acceptability;
(5) Ability to use urea as nitrogen source;
(6) Stability, including resistance to contaminants and inhibitors; and
(7) Clean growth with a minimum of foam.

Drs. Denzel Dyer and Robert Gafford of the Space Biotechnology Section of the Martin Company, have identified a bluegreen microalgae *Synechococcus lividus* that more than satisfies the above criteria. Indeed, it is able to grow at up to 60°C on dilute human urine alone in the presence of light. The accompanying graph (Fig. 32) attests to its excellent nutritional properties when fed to mice.

FOOD FROM FUNGI

Let us turn to the last of microorganisms available as potential SCP: the fungi. Consumption of some types of fungi by man is not new. The cultivated and wild mushrooms and their more exotic cousin, the truffle, are prized additions to the gourmet menu. However, a strict interpretation of SCP could rule out the inclusion of these macroscopic fungi. In Table 4 you can readily see the results of extensive studies with fungal protein.

As I mentioned earlier, filamentous fungi pose their own peculiar disadvantages, owing to their structural nature. Nevertheless, a case can be made for their use in a number of instances where particular substrates lend themselves to fungal growth and where a fungal species itself is a desirable end product. An outstanding example of the latter case is *Morchella*

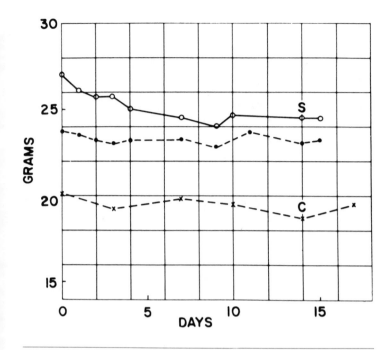

32. Space Age Nutrition
Mean weights of groups of four mice fed *Synechococcus*(S),
Chlorella(C), and standard laboratory ration.
Reprinted, by permission, from Denzel Dyer and Robert Gafford,
"The Use of *Synechococcus lividus* in Photosynthetic Gas Ex-
changers," *Developments in Industrial Microbiology* 3 (New York:
Plenum Press, 1962): 92.

hortensis, a species of edible mushroom. Mushrooms are expensive and perishable and usually grow on some form of compost. The recognizable form of mushroom is the fruiting body and needs no microscope to be seen. John Litchfield and his coworkers at Battelle Memorial Institute, have succeeded in growing the mycelial (see prologue) form in submerged culture yielding an SCP-like product which is not only rich in protein, but has the added advantage of tasting like mushrooms.

A most promising group of Fungi Imperfecti (see Appendix) have been screened by William Gray and his group first at Ohio State University, and then at Southern Illinois.[2] Although the protein yields are not nearly as high as those obtained in algae and bacterial cultures, they have convincingly demonstrated that poor quality or quantity sources of protein such as rice, corn or cassava can be used to increase these yields (Fig. 33). Dr. Gray claims a bonus. His fungi, in contrast to many other SCP's, are tasteless and odorless.

I have by now reviewed many new protein sources, most of which have been grown on otherwise nutritionally deficient substrates. They appear to promise that in the future no one on Earth need suffer from protein deficiency. What are the drawbacks preventing acceptability? If all technical problems were solved and high quality food, i.e., complete in protein, fat, carbohydrates as well as vitamins, could be economically grown (Table 5) and distributed, it still would be necessary to "sell" the product. Cultural and behavioral patterns are difficult for a people to change even in the face of starvation. Who can say how many generations it took man to accept cooked food. Eating habits established in religious and ethnic groups are deeply ingrained. Perhaps the real solution requires the imagination of Madison Avenue in convincing people that their SCP patty is chopped sirloin. Texturing and flavoring has performed miracles for soybean protein, hopefully it can do the same for protein from microbes.

TABLE 4
Results of Feeding Studies with Selected Fungi as Protein Sources

Organism	Form	Species	Per Cent of Dietary Protein	Results
Aspergillus fischeri	Mycelium	Rat	100	Not adequate for growth or maintenance.
A. nidulans	Mycelium	Rat	100	Poor growth; improved with 0.25 per cent cystine or methionine.
A. oryzae	Mycelium	Rat	100	Deficiencies evident; improved with 0.25 per cent cystine or methionine.
A. sydowi	Mycelium	Rat	100	Not adequate for growth. Toxic at 50 per cent of total diet.
Agaricus campestris	Fruiting bodies	Rat	100	Growth less than with casein or soybean meal. Good growth and digestibility.
Boletus edulis	Fruiting bodies	Man	100	Equivalent to muscle protein.
Cantharellus cibarius	Fruiting bodies	Man	100	Equivalent to muscle protein.
Fusarium lini	Mycelium	Mouse	100	Growth comparable to 18 per cent casein ration when ration was supplemented with thiamine.
Heterocephalum aurantiacum	Mycelium	Mouse	100	Adult mice maintained their weight.
Linderina pennispora	Mycelium	Rat	70	Poor acceptance and weight gain.
Morchella esculenta	Fruiting bodies	Man	100	Equivalent to muscle protein.
Penicillium flavo-glaucum	Mycelium	Rat	100	Slight growth at 9 per cent in diet, improved at 18 per cent, or with 0.25 per cent cystine or methionine.
Penicillium sp.	Mycelium	Rat	100	Slow growth, improved with 0.25 per cent cystine or methionine.

*SOURCE: "The Production of Fungi," *Single Cell Proteins,* ed. Richard I. Mateles and Steven R. Tannenbaum (Cambridge, Mass.: M.I.T. Press, 1968), p. 322.

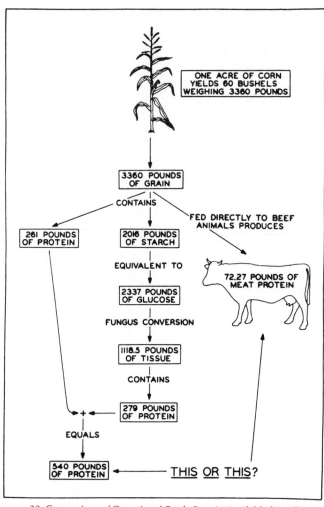

ONE ACRE OF CORN
YIELDS 60 BUSHELS
WEIGHING 3360 POUNDS

3360 POUNDS
OF GRAIN

CONTAINS

FED DIRECTLY TO BEEF
ANIMALS PRODUCES

261 POUNDS
OF PROTEIN

2016 POUNDS
OF STARCH

EQUIVALENT TO

2337 POUNDS
OF GLUCOSE

72.27 POUNDS OF
MEAT PROTEIN

FUNGUS CONVERSION

1118.5 POUNDS
OF TISSUE

CONTAINS

279 POUNDS
OF PROTEIN

+

EQUALS

540 POUNDS
OF PROTEIN

THIS OR THIS?

33. Comparison of Quantity of Crude Protein Available from One
Acre of Corn*
*When fed directly to beef animals and when the corn carbohydrate
is used in the synthesis of fungus protein.
Reprinted, by permission, from William Gray, "Microbial Protein for
the Space Age," *Developments in Industrial Microbiology* 3 (New
York: Plenum Press, 1962): 70.

TABLE 5

Comparison of Anticipated Prices of Selected Animal, Plant, and Microbial Proteins

	Approximate Protein Content %	Approximate Price $ per Pound	Approximate Price $ per Pound of Protein	State of Development
Animal Proteins				
Ocean perch	18	0.54	3.00	Commercial
Hamburger	20	0.54	2.70	Commercial
Dried skim milk	36	0.24	0.67	Commercial
Fish protein concentrate	75-80	0.36-0.48	0.48-0.60	Semi-Commercial
Plant Proteins				
Soy flour	44-47	0.07-0.11	0.16-0.23	Commercial
Soy concentrate	70	0.20	0.29	Commerical
Soy isolate	95-97	0.35	0.36-0.37	Commercial
Cottonseed flour	55-59	0.11	0.19-0.20	Commercial
Wheat gluten	80	0.30	0.38	Commercial
Peanut four	50-60	0.12-0.15	0.24-0.25	Research
Microbial Proteins				
Yeast				
Candida utilis	50	0.15	0.30	Commercial
Candida lipolytica	50	0.15	0.30	Development
Algae				
Chlorella Scenedesmus	45	0.20	0.45	Research
Bacteria				
Methane-utilizing	50	0.15	0.30	Research
Paraffin-utilizing	50-70	0.18-0.25	0.36	Development

*SOURCE: E.S. Lipinsky, I.L. Kinne and J.H. Litchfield, "Technical-Economic Comparison of Microbial, Animal, and Plant Proteins for Food and Feed Uses," *Developments in Industrial Microbiology* 10(1969): 119.

I close this chapter with a quotation borrowed from Henry Peppler.[3] He paraphrases an old proverb, "An almsgiver throws a starving man a fish, but a thoughtful man would give him a hook and line," to read, "A kindly neighbor gives the hungry man a sack of yeast, but a thoughtful neighbor provides him with a single-cell protein starter kit."

6
THE INDUSTRIOUS MICROBE

Unwittingly or not, man has used microbial helpmates in manufacturing processes since the dawn of time. In this chapter I will familiarize you with the scope and variety of useful products made possible, simpler, or more economical due to microbial intervention. This is exclusive of the foods consumed by man, which were covered in earlier chapters.

CLOTHING

Perhaps the earliest products were necessities that man had learned to construct from the natural materials at his disposal. Certainly in less tropical climates one of his first concerns was clothing. Lost in antiquity is that point in time when ancient man learned how to treat animal skins to convert them to leather. This process, called tanning in reference to the use of tree bark extracts or tannins, was not possible without microbes. It is fitting that Louis Pasteur, the genius who gave our civilization the first real insight into microbial activities, should have been brought up in a tannery. His father, Jean Joseph Pasteur, was a tanner by trade and he kept his soaking vats in the cellar of the family house in Dôle, France where Louis was born and spent his early years.

The process of tanning is primarily a treatment of the animal hide to make it less susceptible to (microbial) deterioration. This involves removal of underlying, attached connective tissue and the softening of the skin to allow the better penetration of the tanning chemical, and to make the finished leather more pliable. Heavy salting of the hides removes the attached blood and protein clinging to the underside. The hides are then soaked, or as it is called in the trade, "bated." It is this bating which softens the leather.

In ancient times, bating was accomplished by the metabolic activity of microbes in dog feces added to the soaking hides. Unfortunately, growth of these organisms was a two-edged sword. In addition to conveniently softening the hides, they also scarred and pitted them. Assuredly this was of little concern to our ancestors; however, in the world of today, with perfection our goal in every aspect of life, we are not satisfied with defects. Thus, the uncontrolled bating of hides by resident microbes is no longer tolerated. Instead, chemical products made by microbes are used; these by-products of enzymes will be discussed later.

Microbial action has also been responsible for another step in leather formation: the loosening of hair for easy removal. This dehairing or sweating, as it is known, also produces undesirable side effects. Presently purified enzymes, from both bacteria and fungi, are being used for that purpose.

"And he brought fine linen, and took him down and wrapped him in the linen." Mark 15:46. Linen as a cloth preceded the Biblical period and had been used by the ancient Egyptians for mummy wrappings. It is woven from fibers of the flax plant. In order to get fibers free from the rest of the plant, the flax stalks undergo a process known as *retting*. The stalks are immersed in water and are weighted down, subsequently swelling and also losing soluble carbohydrates. Aerobic bacteria on the plant use up the carbohydrate and at the same time consume the available oxygen in the water. At this time, the

anaerobic species, *Clostridium felsineum* and *Clostridium pectinovorum,*go to work and digest away the polysaccharide matrix binding the flax fibers. This polysaccharide is called pectin and the name *pectinovorum* literally means "eater of pectin." The flax fibers can now be removed, washed, dried, and prepared for spinning. Hemp fiber for rope is prepared in the same way from the hemp plant.

SILAGE
The silo is an everpresent structure dotting this country's farmlands. These buildings are used to store surplus crops for winter feeding of livestock. The fodder (any crop will do—corn, sorghum, potatoes, legume—as long as sufficient carbohydrate is available to produce acid for preservation) is tightly packed into the silo with minimum circulation of air. During storage, the indigenous microbes (as many as 10 billion/ounce), ferment the carbohydrate available to them. This fermentation, called *ensilage,* does little to dimish the nutritive properties of the fodder, gives it an agreeable flavor and aroma, prevents spoilage, and probably produces "contented cows." (See Figure 35.)

TURN ABOUT IS FAIR PLAY
When I arrived in Madrid in 1964 with my family to begin a sabbatical year, we stayed at an apartment-hotel located on *calle* Dr. Waksman. The school my children attended in Madrid was situated on *calle* Dr. Fleming. What a coincidence that I, a microbiologist in Madrid, should find myself on streets named in honor of two Nobel laureates for their work in microbiology. Alexander Fleming was an English physician who in 1928 made the observation that growth of a mold, *Penicillium notatum,* inhibited the growth of the bacterium *Staphylococcus aureus* (Fig. 34). His chance discovery led to the isolation and purification of penicillin in 1940. The story of Selman Waksman is different. He was a soil microbiologist who had been looking

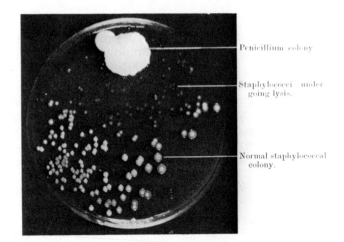

Penicillium colony

Staphylococci undergoing lysis.

Normal staphylococcal colony.

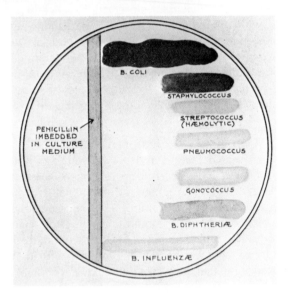

PENICILLIN IMBEDDED IN CULTURE MEDIUM

B. COLI

STAPHYLOCOCCUS

STREPTOCOCCUS (HÆMOLYTIC)

PNEUMOCOCCUS

GONOCOCCUS

B. DIPHTHERIÆ

B. INFLUENZÆ

34. 1928: Alexander Fleming Observed that Growth of a Mold Inhibited the Growth of a Bacterium
Reprinted, by permission, from the *British Journal of Experimental Pathology* 10: 228.

35. The Silo

diligently for antimicrobial agents from soil microorganisms for a number of years. Finally, in 1943, Streptomycin was found as a product of *Streptomyces griseus*. Penicillin and Streptomycin were the first clinically successful of a long line of chemical compounds called *antibiotics,* which seem to be nature's redress for infectious disease. Antibiotics are chemical entities produced metabolically by some microorganisms which in very small amounts kill or prevent the growth of other microorganisms. This definition rules out microbial toxins that affect higher plants or animals and such chemicals as alcohols which are antimicrobial but not at low concentration. It would not, however, rule out those compounds now made chemically but originally isolated and described from microbial origins. An outstanding example of this is Chloramphenicol, discovered in cultures of *Streptomyces venezuelae* by microbiologists at Parke, Davis and Company and known popularly as Chloromycetin, a trademarked product. It was produced commercially from cultures but after its successful chemical synthesis, economics dictated discontinuing biological production.

The purposeful search for antibiotic-producing microbes is a tedious and, more often than not, an unrewarding pursuit. I would estimate that millions of soil samples from all over the world have been examined for the presence of antibiotic-producing microbes. This painstaking work has been almost exclusively the province of large pharmaceutical companies. Soil samples are screened to see if they inhibit any of a large number of selected disease organisms. Once an antibiotic-producing mirobe is isolated and identified, the researcher must answer several important questions:

(1) If the microbe has been described previously, is the antibiotic a new one or the same produced by the earlier isolated microbe?

(2) If the microbe has not been described previously, can you be sure that the antibiotic itself is a new one?

(3) Is the activity of the antibiotic sufficiently high under controlled growth conditions to be practically useful?

(4) Is the animal toxicity sufficiently low (or absent) so that clinical use is warranted?

Obviously, some of these questions are not easy to answer and, in fact, even if the answers were all positive it would not necessarily mean the substance would be produced. If a new antibiotic were found whose range and deficiencies were the same as an existing antibiotic, company marketing research may deem its exploitation economically unsound.

Notwithstanding the above problems, the last 35 years have been a golden age in the chemotherapy of infectious disease, with some thirty different types of antibiotics now available. (see Appendix). With only one notable exception, all these antibiotics are produced by three genera of microbes. *Penicillium* (a mold), *Streptomyces* (an actinomycete, technically a bacterium), and *Bacillus* (a bacterium). There is no "universal" antibiotic, that is, one effective against all types of disease organisms, that can be safely used. Perhaps this is desirable since selective action against a target disease organism is less likely to upset the microbial ecology of the sick host. A case in point: the use of so-called broad-spectrum antibiotics results not in only the death of the infecting microbe but also large numbers of normal flora in the gut and in the vaginal tract.[2] Those killed were dominant bacteria; they were succeeded by a species of fungus (*Candida* or *Monilia albicans*) which results in a nasty infection, moniliasis.

Despite the variety of structures among antibiotics, their modes of action all involve interference with the manufacture of protein, nucleic acid, or cell walls and the function of cell membranes. For example, Mitomycin C, Actinomycin D, and Puromycin inhibit deoxyribonucleic acid (DNA), ribonucleic acid (RNA) and protein manufacture, respectively, in mammalian as well as microbial cells. Thus, they cannot be used clinically to treat disease. However, they are excellent chemical probes for studying the biosynthetic pathways of the above macromolecules.

One of the earliest discovered antibiotics, *Tyrothricin,* is an example of an antibiotic of limited usefulness because of its mode of action. Tyrothricin is a product of *Bacillus brevis,* isolated from New Jersey soil in 1939 by Dr. René Dubos, one of our most eminent microbiologists. It is highly effective against a group of bacteria called gram-positive and acts by attacking their cell membranes. Unfortunately, in the blood stream it would do the same to red blood cells, causing them to dissolve. Consequently, tyrothricin is limited in use to application on the skin where there is minimal contact with the blood.

Sometimes an antibiotic is extremely effective against a large group of organisms, including some that no other antibiotic can handle. Chloramphenicol is such an antibiotic; it is the preferred drug for treatment of typhoid fever (caused by a bacterium *Salmonella typhi*) and typhus fever (caused by *Rickettsia prowazekii,* a flea-borne organism intermediate between bacteria and viruses). However, because it has been implicated in several cases of anemia and leukemia it is now limited to those infectious diseases where nothing else works.

THE MIRACLE MOLD

Some general comments about penicillin. The various penicillins act by interfering with synthesis of the cell wall in sensitive microbes. The growing cell, without the protection of a wall, bursts. Theoretically, these compounds should be safe for animals, and they are. However, in answering a number of questions about penicillin action and therapy, I hope to also straighten out some generally held misconceptions about antibiotics in general.

What is meant by the terms, sensitivity and resistance? I have often heard people say, "I'm resistant to penicillin." They either mean that the microbes infecting them are resistant to penicillin or that they themselves are sensitive to penicillin. Let me explain this.

Resistance in microorganisms we now understand to be a result of "survival of the fittest." In every large population, spontaneous or natural genetic change produces mutants different from their ancestors. The frequency of such an occurence is rare, about one in every ten million genes. One such change might be resistance to the action of penicillin. Normally, these mutants are vastly outnumbered and are not detectable. When penicillin is used as a selective environment, however, it kills all but the mutant, which since it now has the field to itself multiplies and constitutes the next entire population. The new strain of a previously susceptible species has now become the resistant species. Through the years, abuse and misuse of penicillin has created a number of resistant "monster" species, the most notable being *Staphyloccus aureus*, the cause of numerous skin and respiratory infections, and *Neisseria gonorrhea*, the gonorrhea-causing organism.

What about the sensitive patient? A number of individuals, particularly older ones, exhibit a reaction to penicillin. Certain chemical structures in some people called hypersensitive produce sensitivities or (more specifically) allergies. A portion of the penicillin molecule is hyperallergenic, that is, it produces allergies. Thus, the peculiar danger with penicillin is no different than with ragweed. An allergy develops in those sensitive people because of the ubiquity of the *Penicillium* spores in the air and their long-range contact with them. So, because of selective resistance in microbes, allergies in people, and the fact that for many infectious diseases (viral, fungal, and protozoal) no adequate antibiotics now exist—the search for newer and better antibiotics continues.

A closing word about nonmedical uses. Antibiotics have been used successfully to improve the vigor and growth rate of livestock and as a preservative in food. Society has recently begun investigating the potential dangers of antibiotic residues in our food.[3] Governmental restrictions will soon place severe restrictions on such residues, thus making continued use of antibiotics in those areas highly improbable.

ENZYMES

As you may know, enzymes are biological substances that alter the rate of a chemical reation but are not consumed in the reaction. Enzymes, which are protein in nature, make possible most of the metabolic reactions carried out by living cells. Ever since the emergence of microbiology as a recognized science, the production of microbial enzymes for commercial exploitation has been an important industry. Today, well over two dozen types of different microbial enzymes are on the market and some others are in the research and development stages. These enzymes are used in pure research, for medical application, in food technology, in detergents, and perhaps other miscellaneous industrial processes (Table 6). In foods and beverages, enzymes of microbial origin are not only used in the manufacturing processes but are also available for direct use by the consuming public.

A property of all enzymes to a greater or lesser degree is their sensitivity to heat. This means that most enzyme activity is either destroyed or greatly reduced by the cooking or sterilization process. Most of the important applications of enzymes to foods are what the biochemist refers to as hydrolytic, that is, the large molecules, such as starch, proteins, and fats are degraded (hydrolyzed) to smaller molecules such as sugars, amino acids, and fatty acids. The term *hydrolytic* refers to the addition of the components of water (H_2O), H and OH, to the chemical bonds binding the small units into one large molecule, bonds that were formed by removing water originally.

In Chapter II, the brewing of beer was discussed without detailing the events leading from fermentation to bottling. During the maturation period at 32°F, there is a tendency for beer to become hazy due to the presence of undigested grain proteins. The addition of bacterial proteases prior to low temperature storage digests the residual protein, thereby preventing haze. This process is called chill-proofing. Proteases are also used as meat tenderizers, and although enzymes from plant sources (papain from papaya melon and bromelin[4] from pineapple) have had the major share of this market, microbial enzymes are catching up. The use of meat tenderizers by the housewife does not always yield predictable results. Why not tenderize the meat before it gets to the housewife? Preliminary trials of injecting protease into cattle eight minutes before slaughter to allow it to circulate appears to be successful in tenderizing on the hoof.

Rennet is another bacterial protease in increasing use. You may recall in Chapter IV that rennet is used in cheese manufacture for the preparation of the curd. Calf stomach rennet (which has certain objectionable attributes, including the fact that its use violates Jewish dietary laws[5]), is today being supplemented by rennet from *Bacillus cereus* and *Mucor pusillus* (a mold).

Medically, microbial enzymes are used either therapeutically, that is for treatment of some disease condition, or diagnostically, to detect a specific disease.

The most venerable of all enzyme remedies, Takadiastase, was introduced at the turn of the century in Japan as a digestive aid. It is a mixture of starch- and protein-splitting enzymes produced by *Aspergillus oryzae*, the same microbe used in many oriental food preparations (see Chapter V). I will not vouch for its efficacy, but perhaps the best test is its longevity. The recent introduction of a purified starch-digesting enzyme, an amylase from *Aspergillus oryzae,* is now an ingredient in toothpaste. In 1967, a controlled study with an experimental dentrifrice containing a protease and amylase mixture from *Bacillus subtilis* reported a 54 percent decrease in plaque formation, now thought by dentists to be the forerunner of tooth decay.

In the medical arena, two enzymes associated with invasiveness of disease-producing *streptococci* have performed yeoman service in clinical use.[6] They are streptokinase and streptodornase (known by the trade name, Varidase). The former is a protease which activates plasminogen in plasma to plasmin, which is itself an enzyme that dissolves blood clots; the latter directly digests DNA, especially from dead pus cells. These enzymes are very useful for removing clotted blood and fibrinous or pussy exudates resulting from inflammation or in-

TABLE 6
Some Commercially Available Microbial Enzymes

Name of Enzyme	Organisms Involved	Applications
Takadiastase	Aspergillus oryzae	Digestive aid, supplement to bread. Syrup.
Amylase	Bacillus subtilis	Desizing textiles. Syrup. Alcohol fermentation industry. Glucose production
Acid-resistant amylase	Aspergillus niger	Digestive aid
Amyloglucosidase	Rhizopus niveus	Glucose production
	A. niger	
	Endomycopsis fibuliger	
Invertase	Saccharinyies cerevisiae	Confectionaries, to prevent crystallization of sugar. Chocolate. High-test molasses.
Pectinase	Sclerotina libertina	Increase yield and for clarifying juice.
	Coniothyrium diplodiella	Removal of pectin.
	A. oryzae	Green coffee processing
	A. niger	
	A. flavus	
Protease	A. oryzae	Flavoring of sake. Haze removal in sake.
Protease	A. niger	Feed, digestive aid.
Portease	B. subtilis	Removal of gelatin from film (recovery of silver). Fish solubles. Meat tenderizer.
Alkaline Protease	B. subtilis	In laundry detergents.
	B. lichenoformis	
	B. amyloliquefaciens	
Protease	Streptomyces griseus	Same as "Protease—B. subtilis" above.
Streptodornase	Streptococcus sp.	Medical use.
Streptokinase	Streptococcus sp.	
Collagenase	Clostridium histolytium	Medical use. Removal of dead tissue from burns.

Name of Enzyme	Organisms Involved	Applications
Penicillinase	B. subtilis	Removal of penicillin.
	B. cereus	
Glucose oxidase	A. niger	For removal of oxygen or glucose from various foods. Dried egg manufacture. For glucose determination.
Hyaluronidase	P. chrysogenum	Medical use (see text).
Lipase	Various Bacteria	Digestive aid. Flavoring of milk products.
Catalase	Rhizopus	Sterilization of milk.
Keratinase	Streptomyces fradiae	Removal of hair from hides.
Microbial rennet	Mucor pusillae	Cheese manufacture.
Nariginase	A. niger	Removal of bitter taste from citrus juice.
Hesperidinase		
Glucose isomerase	Lactobacillus brevis	Glucose → Fructose.
Laccase	Cariolus versicolor	Drying of lacquer.
Cellulase	Tricoderma koningi	Digestive aid.
Galactose oxidase	Dactylum dendrides	Medical use (see text).
Asparaginase	Escherichia coli	Medical use (see text).

*Enzymes are used in pure research, for medical application, in food technology, detergents, and other miscellaneous industrial processes.

jury. A subcutaneous injection can dramatically remove the cosmetic disfigurement of a black eye much faster than the traditional beefsteak. However, it is for cardiovascular disease that streptokinase is most useful. In acute heart attacks due to a blood clot, it accounts for 50 percent reduction in mortality and in the case of clots in blood vessels or thrombophlebitis, the existing clots dissolve during enzyme therapy.

Another intriguing enzyme is hyaluronidase. Duran-Reynals described it in 1929 as the "spreading factor." Injected subcutaneously in a white rabbit it facilitated the spread of india ink under the skin. The substrate on which it acts is hyaluronic acid, the glue that binds cells together. In some microbial species (Staphylococci), the production of hyaluronidase is associated with the invasiveness of the producing organism. Clinically, hyaluronidase, which is marketed as Alidase and Wydase, is used to facilitate the absorption of fluids injected under the skin, particularly in young children and infants where intravenous feeding is a problem.

In 1964 temporary cures in several cases of lymphatic leukemia were reported after treatment with the enzyme asparaginase. The leukemic cells were only indirectly affected by the enzyme. Apparently, these cells have a greater need than normal cells for the amino acid asparagine, which is destroyed by the enzyme. Although relief is not permanent, any differences between normal and cancerous cells that can be exploited offer hope for other breakthroughs. Presently, production of asparaginase in useful amounts is tedious and costly. Escherichia coli, the bacterium of our large intestine, has been the prime source, but more recently has been joined by Erwinia carotovora which is, of all things, a plant pathogen that rots carrots! Help comes from unexpected places.

The diagnostic laboratory also relies on microbial enzymes. Glucose oxidase converts glucose to gluconic acid. This reaction is conveniently adapted to measuring glucose levels in urine. Another closely related sugar, galactose, can also be tested with its enzyme, galactose oxidase. The early detection of galactose in the urine of infants suspected of having the genetic disease, galactosemia, may prevent more serious consequences such as mental retardation. The basis of this condition is that individuals with certain genetic makeup are deficient in the ability to utilize galactose, one of the sugars resulting from the breakdown of lactose, or milk sugar. Galactose builds up in the blood (galactosemia) and spills over into the urine where it is detected and monitored by its specific enzyme. Phenylketonuria (PKU) is still another disease from an inborn error in metabolism that can be readily diagnosed with an appropriate microbial culture (not technically an enzyme). The growth of a *Bacillus subtilis* is proportional to the amount of the amino acid Phenylanine in urine or blood. (The growth medium which is supplemented by the urine or blood contains a chemical similar in structure to phenylalanine but antagonistic to it.)

Turning to nonmedical uses, the topic of enzymes in detergents is of importance since there has been some disagreement as to the overall effectiveness of microbial enzymes in detergents. The use of enzymes in detergents was first introduced by Dr. Otto Rohm, in 1913, in Germany. Although it was used for essentially 50 years in the European market, it was not a very satisfactory product. The enzyme was of animal origin and was not very stable or active in the alkaline conditions of the detergent.

In 1958 Novo Industri A/S of Copenhagen introduced a protease from a *Bacillus* species that was very active in alkaline conditions. Soon other similar enzymes appeared on the scene. These enzymes are apparently effective in more thoroughly removing stains of a protein nature, such as milk, eggs and blood. They have been added to detergents at about the 1 percent level, meaning 1 part of 100 parts of detergent. As of 1970 they were in over half of the detergents which have an annual market of $810 million; they are responsible for an $80 million new market, the presoaks.

There are two areas of concern in the use of enzymes in detergents. (1) Are they an improvement over the product without the enzymes? All of the consumer panel testing by the major manufacturers say yes; all of the reflectance tests measuring brightness objectively say yes. Notwithstanding consumer acceptance and the above data, question are still being raised by consumer groups and the Federal Trade Commission. (2) Are they safe? Here, our attention is not only with the safety of the eventual user, the consumer, but also the occupational hazard to the worker in the manufacturing plant. There are no indications that the consumer is affected, however there have been reports of dermititis and asthma-like attacks among workers. Currently, with improved equipment for controlling in-plant dust in order to lower available contamination, industry hopes to eliminate these untoward hazards.

VITAMINS, HORMONES, AND AMINO ACIDS

A large number of medicinally and nutritionally important compounds are either synthesized partially or completely by microbes. In some cases, microbial manufacture is the only source. Although all of the B-complex vitamins are synthesized by a number of species of microbes, only Vitamin B_{12} (cobalamin) and B_2 (riboflavin) are of commercial significance. In 1961, sales of riboflavin in the United States exceeded $5 million. This vitamin is not only bought in multivitamin tablets and food supplements, but it is equally important as a growth additive in animal feed.

Vitamin B_{12} is more recent. It was discovered in the late 1940's as a chemical fraction found in the liver that relieved some of the symptoms of pernicious anemia. For example, a dose as low as 10 micrograms (1/3000 of an ounce) restores the red blood cells to normal level. The maintenance dose is even lower. However, although the annual production of B_{12} is only 1/500 of B_2, its commercial value is almost double.

Perhaps, nowhere have microbes been more ingenious than as the organic chemist's alter ego in the transformation of certain hormones known as steroid compounds. The sex hormones and the antiinflammatory hormones belong to this class of compounds. At least in one instance, the conversion of progesterone to 11-hydroxyprogesterone through the addition of a mold, *Rhizopus arrhizus,* helped the chemist leapfrog an extremely difficult organic reaction. (see Fig. 36, in which you will note the "minor" change of the addition of an "OH.") Such microbial intervention has made production of many hormones, such as prednisolone, commercially feasible (Fig. 37).

Glutamic acid is the first amino acid produced in large scale by fermentation: almost 95 percent of the annual world yield (about 220 million pounds) is of microbial origin. Most of this is made in the Orient, ending up as the flavor enhancer MSG (monosodium glutamate). A number of microbes have been utilized in this synthesis, which is essentially a two-step process. Using a suitable carbohydrate as starting material, the following sequences occur:

(1) Carbohydrates and   *Pseudomonas*   Alpha-keto
    ammonia nitrogen +   *fluorescens*   = glutaric acid

(2) Alpha-Keto glutaric   *Escherichia*   = Glutamic
    acid (suitable   +   *coli*   acid
    reaction mixture)

or

Carbohydrates and   *Corynebacterium*   Glutamic
ammonia nitrogen +   *glutamicum*   = acid

Although microbial synthesis of all natural amino acids is accomplished by numerous species, the only processes of interest here are those that are both valuable and economi-

Progesterone + Rhizopus arrhizus = a-Hydroxyprogesterone

36. The Addition of 11-OH to Progesterone by a mold *Rhizopus arrhizus*

Cortisol + Corynebacterium simplex = Prednisolone

37. The Removal of Hydrogen from Cortisol

cally advantageous. Some of the so-called essential amino acids fall into this category, particularly those needed in man's diet because he cannot make them himself. So-called poor proteins are those which are deficient in essential amino acids (see Chapter V). Lysine is the only other amino acid besides glutamic acid to be produced in a large scale. Like glutamic acid it is made by a two-step reaction in which one of the significant organisms is *Escherichia coli*. The most exciting use of compounds like lysine is in supplementing vegetable protein. A leading nutritionist estimates that it would only cost fifty-six cents a year for enough lysine to supplement a corn-and-sorghum diet that would satisfy the caloric requirements of a thirty-pound child. This is a small price to pay to prevent kwashiorkor.

CHEMICAL MISCELLANY

I cannot close this chapter without some mention of the metabolic products of the mold, *Claviceps purpurea*. This organism is a parasite of rye in nature and if its products get into rye flour, it causes what is known as "ergot poisoning." The first description of ergot poisoning appeared in the Middle Ages, although it was not until 1670 that the poisonings were actually attributed to the consuming of contaminated grain. The descriptions of the epidemics were rather horrifying—tissues of the arms and legs became dry and black with gangrene and in some cases the limbs would actually fall off. Limbs were said to be consumed by the Holy Fire, and because of this the disease was called Holy Fire, or St. Anthony's Fire, the latter name honoring St. Anthony at whose shrine many ergotics recovered. (This was probably true because of their change of diet during the pilgrimage: that is, they were fed bread from uncontaminated grain en route and at the shrine.) But was there a positive side? Ergot has been described as a veritable treasurehouse of pharmacological constituents. Perhaps its most widespread use is as an oxytocic agent, that is one that causes contractions of the uterus. Clinically it is prescribed (1) to induce labor, (2) to

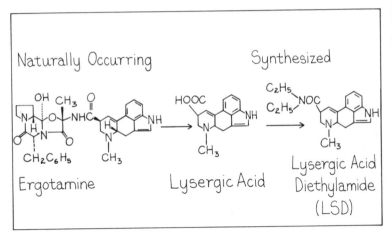

Naturally Occurring   Synthesized

Ergotamine   Lysergic Acid   Lysergic Acid Diethylamide (LSD)

38. Derivatives of Ergot and Their Relationships

control post-delivery hemorrhage, and (3) to help return the uterus to normal following delivery. The other major *legitimate* use is in the treatment of migraine symptoms.

About fifteen years ago, it was discovered that a semi-synthetic derivative of lysergic acid, one of the ergot alkaloids (Fig. 38), in very low doses produced hallucinations, or as it was later called, psychodelia (see Fig. 38). This compound is lysergic acid diethylamide, or LSD. It was originally used for psychological research, but because of unpredictable side effects, was abandoned. On the black market, LSD has led a whole generation on a trip as unpredictable and dangerous as St. Anthony's Fire.

What is left? Literally, a shelf full of all kinds of chemicals that are cheaper and simpler to make microbiologically. Citric acid, that ubiquitous acid of most fruits, is made commercially through the fermentation of glucose by *Aspergillus niger,* and is an additive in a wide variety of foods, such as soft drinks, gelatin desserts, candies, preserves. It is also found in effervescent tablets (e.g., Alka-Seltzer) and the sodium salt of citric acid is the anticoagulant in blood drawn for transfusion.

I am sure that my attempts at presenting the whole area of productive industrial microbiology in one chapter has resulted in some sins of omission. I knowingly omitted, due to space considerations, any discussion of "microbes in mining," which for some minerals are highly effective additions to ore extraction technology. Nevertheless my survey, intended to cover the areas most relevant to everyday needs, should be complete enough to have convinced you of the benevolent potential of microbes.

And what does the future hold?

Microbial genetics are being used in two ways. Under the leadership of Dr. Bruce Ames of the University of California, tests have been devised which relate the mutation of selected bacterial strains by environmental chemicals to their cancer-causing potential, making it possible to screen rapidly (two days as opposed to two years) new chemicals for this type of biological activity.

The microbial geneticist is also utilizing the genetic proclivity and metabolic capacity of the bacterial cell to synthesize proteins originally derived from distant genetic sources. This is the realm of recombinant DNA research, in which the scientist literally inserts bits and pieces of genetic information derived from donors as distant as mammals into a responsive bacterial cell, such as *Escherichia coli*. Under appropriate physiological conditions, the bacterial cell receives the genetic message and translates the information into a protein. This bit of biological engineering is done primarily with the transfer of units called "plasmids," which are extrachromosomal inheritance units.

Because of the fear, however poorly founded it may be, that these experiments could produce genetic monsters, severe restrictions on recombinant DNA research have been imposed at the federal level (*Federal Register* 42:49595–99, September 27, 1977). However, the potential for good with this system should be obvious. Several laboratories have focused their attention on the microbial synthesis of the mammalian hormone insulin, so desperately needed by diabetics, whose only source at present is the pancreas from three species of animals. At this writing, the synthesis by bacteria of the polypeptide human hormone somatostatin has been reported (*Science* 198:1056–63, December 9, 1977). Although the technical difficulties are great, they are gradually being overcome, and the future use of the microbial cell as the biochemical engineer holds unlimited possibilities.

7
THE LIVING END

"In the sweat of thy face shalt thou eat bread, till thou return unto the ground; for out of it wast thou taken: For dust thou art, and unto dust shalt thou return" (Genesis 3:19).

Consider, for a moment, the consequences of the above prophecy *not* coming true. Since the dawn of life we would have been accumulating the detritus of death. How fortunate that natural decay has prevented this from happening! Decay is only a small part of the ongoing process in nature involved in the recycling of all the elements in the biosphere. That recycling is also a reflection of the multitude of associations among microbes, and between microbes and higher biological forms. The intonation by John Donne that no man is an island, no man lives alone, is a truism for all living things in the natural world. This chapter will explore some of those associations and interactions, in particular where the resulting activities are beneficial to man.

We are in the decade of ecological consciousness. Suddenly everyone has become aware of the limits to our natural resources. Pollution and biodegradability have become household words.[1] What is biodegradability? A simple definition would be nature's recycling of elements in complex substances within a reasonable period of time. Although it had been considered almost axiomatic that all oxidizable organic compounds would find a host to dine upon them, all we need do is look around us to find examples of organic materials that have resisted the ravages of biological degradation (Table 7).

TABLE 7
Estimated Age of Nonbiodegradable Compounds

Material	Age in Years
Leather	2000
Humus	35,000
Lignite	1 million
Amber	25 million
Chitin	500 million

Recently much study has gone into finding out the basis for this resistance to biodegradation. Martin Alexander of Cornell University suggested that we discard the principle of microbial infallibility and replace it with a more rational understanding of what he refers to as molecular recalcitrance. In other words, if we examine the reason why a compound or series of compounds resists microbial attack it may be possible to overcome these difficulties. Alternately, it may be possible to design compounds that resist attack while in use, thereby preventing deterioration, but once the compound is discarded proper biological agents and/or environments may be used to inhibit the factors responsible for nonbiodegradability. But that's another story, only part of which will be treated here. Thus, although our present environmental crisis has focused attention on the accumulation of nonbiodegradables, this is neither a fault nor a virtue of the microbe. There are extreme examples of nonbiodegradability (Table 7) that have proven to be providential. These are the apparent indigestible leavings of the carbon cycle, in which microbes play a leading role. I refer to our vast subterranean deposits of the fossil fuels, coal and petroleum, which are the major sources of energy in use today.

THE CARBON CYCLE

Let us examine the carbon cycle. It has been estimated that about one billion metric tons of organic matter are broken down by decay to carbon dioxide and water annually on this planet. If the supply of organic matter were not replenished it would all be gone in less than twenty years. Where does all this energy go? Some of it is consumed in the processes needed to support life, but most of it is uselessly burned biologically and is given off as heat (this is what scientists refer to as an increase in entropy or randomness. A very small part is partially degraded and deposited as coal or oil. This is a dead end and is actually not part of the cycle.

What keeps us going then? Where does the energy come from to maintain life? Although it is difficult to be exact with such figures, several investigators have shown that photosynthesis more than replaces the organic matter lost through decay. One sum given is 54 to 94 billion metric tons of carbon fixed per year. A significant portion of this is produced in the seas of the world by phytoplankton (algae). Thus, the carbon cycle is beholden to microorganisms for releasing CO_2 in decay, and for the subsequent regeneration of organic compounds for the sustenance of more dependent life forms.

However, it would be a gross oversimplification to limit (by implication) the carbon cycle to CO_2 generation and CO_2 fixation in photosynthesis. The subsequent release of CO_2 from dead plant and animal tissue via respiration of microbes is really the result of the concerted action of a large variety of species, each with its own nutritional proclivity. For example, digestion of cellulose is carried out by some fungi as well as bacteria belonging to a rather exotic group, the myxobacteria. In addition, lignin and pectin from plants and fats, proteins and other carbohydrates from plants and from animals must be digested. To these add the mixed bag of chemicals made by man: herbicides, insecticides, and all sorts of plastics, all of which contain carbon. Earlier in the book I mentioned some dead ends. One of these can be methane, or marsh gas, produced by *Methanobacter* and *Methanococcus* species. This is a fermentation reaction. Sometimes in marshy areas, when climatic conditions are just right, the methane spontaneously ignites. The dancing flame of burning methane at night was associated in ancient folklore with the "Will O' the Wisp" or Jack O'Lantern, while in more modern folklore with "Flying Saucers." Under aerobic conditions in the environment methane is oxidized; that is, it serves as an energy source for a group of bacteria known as *Methanomonas,* putting CO_2 back into the atmosphere. What about the return of CO_2 from the atmosphere in the form of fixed carbon? In addition to green plant photosynthesis which fortuitously returns oxygen back to the atmosphere, some bacteria can also fix CO_2 in the presence of light, albeit of a different and noncompetitve wavelength than that used by green plants. They perform heroically in their own ecological niches. There is another nutritional group of bacteria that have no counterpart in the living world. They fix CO_2 with the energy derived not from light but from the oxidation of minerals such as hydrogen sulfide and nitrous acid. Mother nature has many mansions! The carbon cycle emphasizing the role of microbes is summarized in Figure 39.

MICROBIAL DECAY AND THE DEAD ENDS

The nonbiodegradables listed in Table 7 have resisted the ravages of time because of the lack of oxygen, or the absence of moisture. So under optimal conditions in nature even those indigestible substances would not accumulate. Humus and lignin, for example, are both degraded microbiologically in the soil. Humus is the normal organic content of soil and it is estimated that in average tillable soil no more 2 to 5 percent is degraded annually. However, this is quite variable and is the result of many factors, not the least of which is the presence of added organic material in the form of fertilizer. This addition increases the growth rate and total numbers of soil microbes

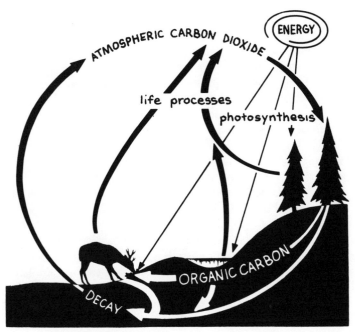

39. Carbon Cycle
Reprinted, by permission, from *High School Biology*, 2d ed., BSCS
Green Version (Rand McNally & Co., 1968), p. 24, courtesy of Biological Sciences Curriculum Study, Colorado.

which subsequently contribute to the increase in humus decomposition.

Cultivation also enhances the rate of humus decay; aeration of the soil plays a large role in this disappearance. Lignin, which may be a major source of humus in soil, is the third most abundant constituent of plants and the most resistant to decomposition. The major organisms involved in this decomposition appear to be fungi and filamentous bacteria. The process is inhibited by low temperatures as well as excessive moisture.

Under minimal conditions of decomposition there is a considerable body of evidence that both humus and lignin are converted to peat and coal, although only the former conversion can be observed to any degree at the present time.

The initial conversions of decaying plant materials to peat and coal are aerobic and involve bacteria and fungi. The secondary, slower transformations into the very highly stable products are anaerobic and are biologically derived in the production of peat, but are probably abiogenic for the coals.

The possibility that petroleum deposits may be derived from microbial sources is quite controversial. However, even though direct evidence is lacking, there is strong support for such an hypothesis. One of its strongest proponents is Dr. Claude ZoBell of the Scripps Institute of Oceanography. He has suggested that the elemental composition of marine humus (Table 8) appears to be intermediate between that found in marine organisms and in oil. He has stated that microorganisms found in oil deposits are normal inabitants and not contaminants brought in from drilling. Laboratory studies with some of these anaerobic species showed them capable of synthesizing liquid hydrocarbons which ZoBell says can be accumulated in sedimentary rocks when the microbe dies. As attractive as this hypothesis is, no noticeable accumulation of oil has been found associated with the ooze deposits of bottom muds where one would suspect that they would begin to appear.

SOIL FERTILITY AND THE NITROGEN CYCLE

The process of decay which I have just discussed in relation to the carbon cycle and fossil fuel formation can be related to what is called mineralization, the process by which complex organic molecules are converted to inorganic compounds and elements. Some of these are available for green plant growth. Thus, soil with relatively large numbers of living microorganisms, in excess of 100 million per gram, is generally more fertile than soil with fewer, a condition common in desert areas. The element most important to fertility is nitrogen and its cycling in nature is no less important than carbon (Fig. 40). Between nitro-

TABLE 8

Composition by elements of the organic matter of oil and of the initial organic matter from which oil may have originated (%)

Element	Marine organisms	Organic matter Marine humus	Oil
Carbon	45–32	52–58	82–87
Hydrogen	5–9	6–10	11–14
Oxygen	25–30	12–20	0.1–5
Nitrogen	9–15	0.8–3	0.1–1.5

*Source: S. I. Kuznetsov, M. V. Ivanov and N. N. Lyalikova. *Introduction to Geological Microbiology,* trans. Paul T. Broneer, Ed. Carl H. Oppenheimer. New York: McGraw-Hill Book Co., Inc., 1963 (p. 82, Table 32).

gen fixation, the uptake of nitrogen gas from the atmosphere, and denitrification, the release of nitrogen back to the atmosphere, there exist a myriad of biochemical reactions in which nitrogen is the keystone element. Let us begin with perhaps the most obvious part of the cycle; organic nitrogen degradation. Dead animal and plant material, as well as animal wastes, undergo a process of digestion and putrefaction by soil bacteria, yielding primarily a mixture of amino acids. These end products are food for protein synthesis in microbial growth so that organic enrichment in the soil increases fertility. These amino acids can also serve as starting substances for ammonia formation (ammonification), the next step in this cycle. Now the beginning of mineralization! Ammonia is oxidized to nitrous acid by species of *Nitrosomonas,* a genus belonging to a nutritional group discussed earlier, encompassing those microbes which derive all their energy from oxidizing minerals. Then the nitrous acid is converted to nitrate by species of *Nitrobacter.* Nitrate is a crossroads compound. It can serve as plant food and be taken up by hungry plants; it can be reduced by certain microbes all the way to gaseous nitrogen, the process known as denitri-

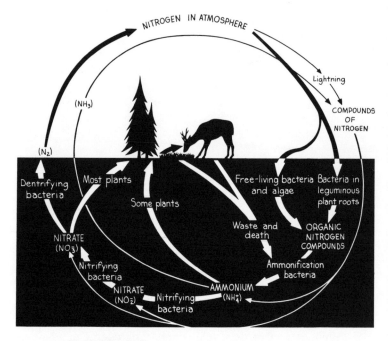

40. Nitrogen Cycle
Reprinted, by permission, from *High School Biology*, 2d ed., BSCS
Green Version (Rand McNally & Co., 1968), p. 243, courtesy of
Biological Sciences Curriculum Study, Colorado.

fication. We can now begin to look at the reverse trip of nitrogen back from the atmosphere. This is called *nitrogen fixation*.
Microbes in soil (and in water, too) alone and in concert with
certain green plants convert nitrogen from the air to a form
which in effect is a fertilizer. The two general types of microorganisms involved are free-living and symbiotic species respectively, the latter being those organisms that work cooperatively
with another species for their mutual benefit. The free-living
species are found in a broad spectrum of groups, including
aerobic and anaerobic and heterotrophic and photosynthetic
bacteria and blue-green algae. Several reports have suggested

that some yeasts and fungi could fix nitrogen, but more recently doubts have been cast on such studies.

Nevertheless, the species that have proven themselves beyond a shadow of a doubt as nitrogen fixers have made a significant contribution to returning nitrogen to the biosphere. The two most commonly investigated in nitrogen fixation studies, *Azotobacter* (a name which means the nitrogen bacterium) *vinelandii* and *Clostridium pasteurianum* are both heterotrophic, that is, they need some form of organic food such as sugars for growth. Since nitrogen fixation only occurs during microbial growth, for every N_2 molecule plucked out of the air an equivalent amount of sugar must be available as energy. It is for this reason that these species play a negligible role in soil fertility in cultivated land, since their contribution there requires organic enrichment to achieve that energy. However, the overall importance of these species cannot be minimized; they are responsible for completing the cycling of elemental nitrogen back to terrestrial environments.

In contrast to the above, there is evidence that in flooded soil nonsymbiotic nitrogen species are important in crop production. In rice paddy fields good yields are produced without organic enrichment. Quite often these flooded paddies are covered with blooms of blue-green algae species capable of fixing nitrogen, needing only carbon dioxide as a source of carbon (meaning they are photosynthetic). Considering the reliance on rice as a major grain crop in parts of the world where nitrogen-rich fertilizer is hard to come by, fixation by blue-green algae and photosynthetic bacteria may indeed be a major contribution to fertility.

You Can't Get Blood From a Turnip—But Would You Believe Soybeans?

A most unique association in the biosphere involves a large group of plants and one genus of bacteria. The plants, the pea and bean family (*Leguminosae*), which also includes peanuts, vetch, alfalfa, and clover, have a symbiotic relationship with the *Rhizobium* species. These bacteria infect the plant roots and in so doing, cause nodules or swellings in these roots (Fig. 41). These nodules are factories where the product is fixed-nitrogen and the raw materials are elemental (atmospheric) nitrogen and sugars made by the plant. Although there are undoubtedly a number of growth factors and nonspecific minerals necessary for Rhizobial growth, the central theme of this mutualism is the bacterium's furnishing of mineral nitrogen to the plant in exchange for energy-rich carbon. The relationship is not absolutely necessary for survival since the plants and the bacteria can live without each other. However, in the absence of nodulation, the legumes need nitrogen fertilizer added to the soil for optimal yields, and the bacteria can grow in test tubes only when sugar and previously fixed nitrogen are available. They cannot fix nitrogen hosts.

The extent of nitrogen fixation by nodulated legumes can average one hundred pounds per acre per year calculated under optimal conditions. Thus, in cultivating these crops, there is no need to add nitrogen-containing chemical fertilizers. And as an added plus: legumes do not decrease soil fertility as do nonlegumes when they are continuously cultivated. When plowed under, crops such as clover or alfalfa markedly enrich the nitrogen content of the soil. However, even when the above-ground portions of the plants are consumed by grazing animals, there is a slight net increase in nitrogen when the manure is added back to the soil. I should point out here that man is in considerable debt to microbes for his dietary protein, not only for their contributions to nitrogen fixation in such edibles as peas and beans, but also the subsequent conversion of grazed materials into steaks, chops, and milk.

Biologically, nodule formation is an inflammation where plant root tissues respond to infection with the Rhizobial species, and increase abnormally in size to form the tumors called nodules. One apparent mechanism for this overgrowth of

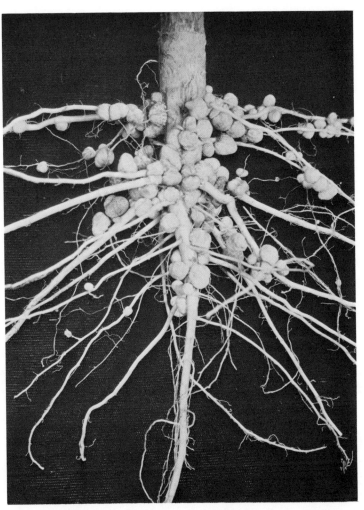

41. Bacteria Infecting Plant Roots and Causing Nodules or Swellings
Soybean roots that have abundant nodules formed with a species of
Rhizobium.
Reprinted, by permission, from *High School Biology,* 2d ed., BSCS
Green Version (Rand McNally & Co., 1968), p. 245. Photograph
courtesy of the Nitragen Company.

the plant cells is the production by the bacteria of a plant growth hormone. The bacteria also multiply during this time. Collectively, the plants and their bacteria produce an unusual protein (unusual, that is, for plants) without which there is no nitrogen fixation. This protein is a hemoglobin similar to that found in human red blood cells and it imparts a pink tinge to otherwise colorless nodules. No doubt if you were to squeeze them hard enough, they indeed would bleed.

ANIMAL ENERGY FACTORIES
"Whatsoever partest the Hoof and is cloven footed, and cheweth their Cud, among the Beasts, that shall ye eat" (Leviticus 11:35). These are the ruminants, and include among them the families of cow, goat, sheep and deer. They have been endowed by evolution and nature with the intrinsic capacity for receiving energy and protein from grazing, that is, plant food. In effect, these animals are factories that convert low-quality plant-protein nitrogen to high-quality animal protein. The workers in this factory, which seemingly labor around the clock, are microorganisms residing in the rumen, the definitive compartment in the multi-partitioned stomach of the ruminants. This labor-management arrangement is one of nature's most highly refined examples of mutualism or symbiosis.

Let us look at the structure of this remarkable energy factory (Fig.42). The ruminant stomach is said to be, perhaps facetiously but with more than one grain of truth, more highly evolved than the human brain. There are four anatomically distinct functional parts: the reticulum, the rumen, the omasum and the abomasum. The rumen, however, is the most important portion as far as this discussion is concerned. The contents of the rumen and the smaller adjacent reticulum can constitute up to one-seventh of the weight of the ruminating animal!

Here's how it works: in the adult grazing animal, food is bitten off (but not chewed), thrust to the back of the mouth by the tongue and literally explosively swallowed. These animals

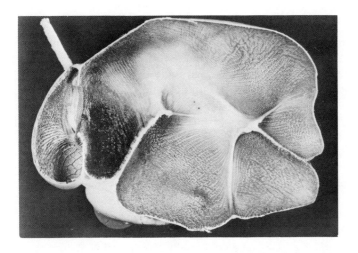

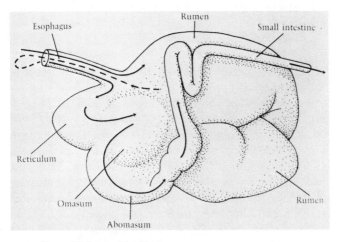

Esophagus
Rumen
Small intestine
Reticulum
Omasum
Abomasum
Rumen

Figure 42. Rumen Nutrition
The rumen and gastrointestinal system of a cow, showing the route
and passage of food.
Diagram reproduced by permission of the publisher from Thomas
Brock, *The Biology of Microorganisms*, 2d ed. (Englewood Cliffs,
N.J.: Prentice-Hall, Inc., 1974), p. 472. Photograph reproduced by
permission of N.J. Benevenga from the *Journal of Dairy Science* 52
(1969): 1294.

are endowed with muscles in the esophagus which are unique in the animal kingdom and which propel the food into the reticulum so forcibly that it ends up in the rumen. Here the forage is mixed with a large amount of saliva secreted by the animal (for cattle it is twenty-five gallons per day) and the microorganisms begin converting the plant food to their own ends. For tender leafy vegetation, microbial enzymes suffice to digest all edible proteins. However, certain coarse herbiage will take an eternity to break down with enzymes. To meet this problem, the ruminant has evolved a convenient physiological mechanism: they vomit and rechew their food. This is called *chewing the cud* and whole process is called *rumination*. Thus, the ruminant animal grinds up its food into small pieces, suspends it in saliva and continually mixes it all with contracting of the rumen musculature. What microbes are involved and what is their fate?

The Workers. Of the more than 30 billion living bacteria per ounce of rumen contents, no physiological group is obviously more important than the cellulose-digesting species. At least a dozen different species have been identified, including both rod- and coccal-shaped types as well as aerobic and anaerobic forms. It is estimated that these types make up only about 5 percent of the total living population. Cellulose digestion is the keystone of ruminant nutrition since it provides energy via microbes from a source not normally available to animals. There are also numerous species utilizing other carbohydrates such as starch and sugars, and although carbohydrate digesters predominate in the rumen there are also protein- and fat-digesting species.

In addition to the varied and numerous species of bacteria, the normal rumen also harbors a mixed bag of protozoa (single-celled animals). These microscopic animals, ranging from 1,000 to 100 million per ounce, depending upon the ruminant and its food, undoubtedly make an important but apparently nonessential contribution to its nutrition. Shortly after birth, the young ruminants are inoculated with protozoa by their mother's grooming. This process is called faunation. By the reverse process, or defaunation, it has been possible to study the effect of the protozoa on ruminant growth. The results have not been conclusive, although recent research in lambs showed a greater weight gain in faunated animals. In the absence of protozoa, their nutritional functions seem to be taken over by bacteria.

How Does the Ruminant Utilize Its Flora? The forage consumed by the ruminant remains in the rumen for as long as 36 hours, depending on its fibrous nature. The undigested portions, along with the microbes that recently dined upon it, pass along to the omasum via the reticulum. The omasum functions primarily as a holding chamber in which soluble nutrients produced in the rumen are absorbed and in which large particles are prevented from proceeding further. After screening and absorption, the food passes into the abomasum which resembles in function the stomach of nonruminants. It secretes hydrochloric acid and pepsin, partially digests microbial cells and discharges its contents into a typical mammalian small intestine. Thus, what essentially serves as food for the weaned ruminant are products of microbial growth and the microbes themselves.

I have purposefully avoided detailing the structure of this remarkable organ, the rumen. It is an engineering complexity, filled with numerous one-way valves and by-passes. One point bears mentioning. In the nursing young, the sucking reflex shuts off the valves to the rumen, permitting the milk to go directly to the abomasum, where it is digested as it would be in nonruminants.

The Future. Although ruminants do exceedingly well on high-protein forage like alfalfa, much of this quality food is not essential for ruminant survival. Indeed, a good bit of it ends up in manure, which is good fertilizer but, in these days of mass feed lots, creates a distinct disposal problem. However, this shrinking planet is slowly running out of space for many things, not the least of which is grazing space for all those steaks and chops so desired by an increasingly affluent society. Partial answers may not be far off. Artturi Virtanen, the Finnish scientist who was awarded a Nobel Prize for his work in nutrition, found that urea and ammonia could completely supplant plant protein in dairy cattle feeding, making possible the utilization of a broader selection of forage. You will recall that in Chapter 5 ("Food for Thought"), I mentioned the potential for single-cell protein might well be in the production of ruminant protein. This, in effect, would offer microbial protein produced outside of the rumen and would reduce the amount of earth space needed for grazing these animals. However, despite the external supply of protein we must still rely on the sophistication of the very complex ruminant digestive tract to convert bacteria to beefsteak.

Our Resident Flora. In the previous paragraphs, I have discussed with you a rather clear-cut case of a necessary, natural interrelationship, that of the ruminants and their microbes. However, all living things in this world coexist with their own array of microorganisms, from the microscopic bacteria with their own viral parasites (the bacteriophage) to the variety of flora and fauna found in and on the giant blue whale. For many years, there has been a great deal of speculation regarding the contribution of so-called normal flora to the well-being of the host. Could animals exist without their microscopic residents? Would they be better off without them? What specific plusses or minuses are involved? The answers to these questions required the development of a new science and with it, new technology. So we have the study of germfree life.

By excluding microbes from the environments of animals at birth, by physical barrier and sterility control, it has been possible to raise mammals as large as sheep and goats to weanling size without subsequent contamination. With obviously lesser difficulties, this also has been possible with birds. Reared this way, these are "germfree" animals. From them we can begin to learn whether we are capable of existing in a biological vacuum. We can begin at a zero baseline and then add one microbial species at a time to the host, in this way learning the specific contribution by that microorganism. Whether an animal is germfree or whether it is harboring one or more species given to it after the establishment of germfree conditions, both have one thing in common: their total microbial population is known.[2]

There are obvious advantages for biological research in knowing the total population. However, such a discussion is not within the aims of this book. Only one question is germane. Are our so-called normal flora (often called commensals) essential for survival? The only studies conducted so far of any relevance for man (and this may be questioned) are those done on rodents. Even here there appear to be differences of opinion. For example, one study indicates that germfree mice have a greater average longevity while another study shows just the opposite. There are rather obvious anatomical differences. The lymph nodes of the germfree animal are extremely small. The swelling of lymph nodes in nongermfree animals is due to the continual stimulation of those tissues as they filter foreign particles (including microbes) from the circulatory system. These nodes are the primary source of cells that make antibody molecules. Thus a swollen lymph node is like a well-exercised muscle; it is more apt to handle the external challenge of external disease microbes.

The large intestines, which normally harbor large numbers of varying species of bacteria, are less developed in the germfree animal (Fig. 43). This offers no particular hardship to the animal in post-germfree existence. However, one clear-

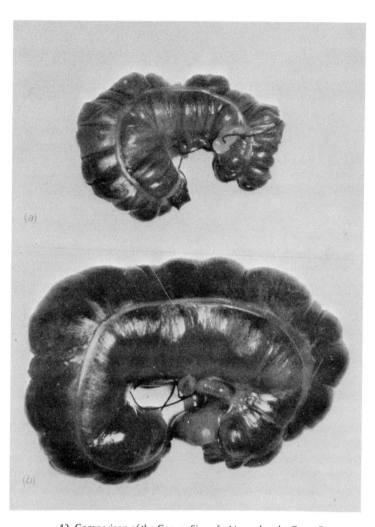

43. Comparison of the Cecum Size of a Normal and a Germ-Free
Rodent
Reprinted, by permission, from Thomas Brock, *The Biology of Micro-
organisms,* 1st ed. (Englewood Cliffs, N.J.: Prentice-Hall, Inc., 1970),
p. 398. Photograph courtesy of Medical Audio-Visual Branch, Walter
Reed Army Institute of Research, Washington, D.C.

cut contribution of intestine flora to rat nutrition has been established. Germfree rats on a diet deficient in Vitamin K show prolonged blot-clotting times. The role of Vitamin K in the biosynthesis of prothrombin in the liver is well known and animals that do not make their own need an external source. Conventional rats need no dietary source of Vitamin K. The conclusion: intestinal microorganisms produce Vitamin K for the rat.

These are all artificial situations and to cite them as reasons for the absolute need for a normal flora would be faulty. However, the normal flora live in harmony with us, prepare us for dangerous invaders, and when left undisturbed prevent the incursion of potentially dangerous species. Thus, even though there is no vast body of scientific data demonstrating the necessity for our microbes that live on us, I am tempted to quote Louis Pasteur, who said, "Life would not long remain possible in the absence of microbes."

WASTE NOT, WANT NOT

Certainly no treatment of the activity of microbes in nature, and in regard to man, can be complete without a discussion of microbial roles in waste and sewage disposal. Man has always had an ambivalence about his own waste, from coprophilia to outright rejection and revulsion. Certainly, the healthy attitude would lie somewhere in between. Waste deposition by many animals is a behavioral characteristic that serves to establish individual territories and to leave markers for companions to follow. When did ancient man develop his taboos about his own excrement, taboos that stretched across the centuries to become one of the cornerstones of Freudian psychiatric theory? When, indeed, did man learn that excrement had real value?

Man is a thinking animal and let us give our ancestors more credit than perhaps they deserve. Maybe he disposed of his waste because he coincidentally associated it with the spread of disease. After all, the revulsion of excrement by modern man because of smell or appearance or whatever, is a learned subjective reaction. The ancient Hebrews committed to writing many public health recommendations that have stood the test of time, including an invocation to dispose of personal waste: "Thou shalt have a place also without the camp, whither thou shalt go forth abroad. And thou shalt have a paddle upon thy weapon; and it shall be, when thou wilt ease thyself abroad, thou shalt dig therewith and shalt turn back and cover that which cometh from thee" (Deuteronomy 23:12-13).

Perhaps lost in antiquity was the husbandman or shepherd who first noticed that the grass was greener over the place he had used as a privy in the recent past. In any event, the sequence that renders excrement harmless and less obnoxious also yields the products that make the greener grass. This is the basis of organic gardening, and a part of the recycling discussed earlier in this chapter and also in Chapter 6. Soil microbes have been hard at work! This is the most primitive type of waste treatment. Microbes make the direct transformation. Many of my readers who have served in the armed forces may remember with some amusement the directions for digging cat holes, slit trenches and the more sophisticated latrine pits. However, more organized structures for waste disposal have been with us for a long time. Ancient Rome had a sewage disposal system, the Cloaca Maxima, perhaps the first engineered facility of its kind in man's history.

Despite his advanced technology, modern man has infinitely more problems with waste disposal than his ancestors. Not only is there a greater variety of waste material needing disposal, but let's face it, we are running out of space for disposition. I partially discussed this problem earlier in this chapter in terms of biodegradation. There are three separate types of waste needing disposal: urban waste water (including domestic and industrial waste), agricultural waste (essentially livestock waste), and solid waste (including garbage and rubbish).

Urban waste water is better known to you as sewage, and to a greater or lesser degree it undergoes treatment in large metropolitan treatment plants. Sewage is more than 99 percent water, and consequently there is not much to remove. The lax attitude we have had in the past about this little bit has resulted in the pollution headaches from which we are now suffering. The aim of waste water treatment is to treat the water so that it can be safely returned to the natural environment. The simplest, crudest method and one not involving microbes to any great extent is called primary treatment. This process involves removal of large objects by screening and so-called settleable solids by detention. The waste water is then chlorinated and discharged back into a natural waterway, such as a river. This process does very little to remove the dissolved organic and oxidizable inorganic matter and almost nothing to remove soluble inorganic matter (phosphates). Secondary treatment processes have been developed which in fact are essentially microbiological processes. These processes are designed to reduce biological demand, or BOD, a term which refers to the amount of oxygen consumed by microorganisms in converting food to CO_2. If waste with a higher BOD is discharged into natural water, oxygen levels may be depleted, thus depriving other aquatic life forms of an adequate supply of oxygen. In principle, BOD reduction is produced by activated sludge (Fig. 44) and/or anaerobic digestion. The former relies on the activity of aerobic bacteria—those that need air to get energy from their food. The settleable and suspended solids removed in primary treatment are referred to as sludge and contain the indigenous microbes. The activated sludge process may be defined as a "system in which flocculated (suspended) biological (mainly bacterial) growths are continuously circulated and contacted with organic waste in the presence of oxygen."[3]

Over a period of time, the microbes in the sludge become acclimatized physiologically to waste water nutrients and make short work of the organic food and oxidizable inorganic minerals in the incoming (influent) sewage. This process of selective adaptation of the sewage microorganisms is what makes "sludge activated." In an efficient activated sludge op-

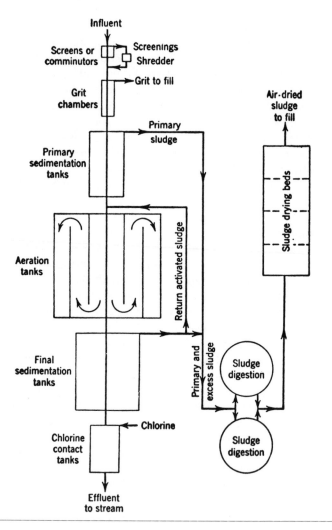

44. Activated Sludge
Schematic flow diagram, complete treatment, activated sludge.
Reprinted, by permission, from Phelps, et al., *Public Health Engineering: A Textbook of the Principles of Environmental Sanitation,*
Vol. 1 (John Wiley & Sons, Inc., 1948), p. 571.

eration, the prepared microbes can remove as much as 90 percent of the BOD in fifteen minutes.

In terms of space and time, this process is preferred to anaerobic digestion, especially in the largest of our metropolitan areas where commonly one billion gallons of waste water daily are produced. The net result of all this activity is production of CO_2 which is released to the atmosphere, a reduction in organic content in the waste water, and a greater sludge volume due to microbial growth.

Because of the space and time problem, anaerobic digestion is rarely used as the exclusive secondary treatment process in large metropolitan treatment plants. Commonly, its use is limited to digesting domestic organic sludge and concentrated industrial water sludge. Average detention times for anaerobic digestion are approximately thirty days. During that time, in the absence of air, sludge is fermented; that is, end products are formed that are characteristically associated with fermentation (see Chapter I). However, the end products are gases like methane, carbon dioxide, and nitrogen, which is derived from compounds such as organic acids and nitrate. Anaerobic digestion is then a useful mechanism for recycling nitrates that may have accumulated in prior activated sludge treatment; this, in effect, is an important plus in some areas of disposal, since the accumulation of nitrate can be a dangerous public health hazard in ground waters. The generation of methane (the same as natural gas) is sufficient in most treatment plants to provide fuel for their heating.

As efficient as the combination of activated sludge and anaerobic digestion is in lowering the contamination level of waste water, soluble nonoxidizable compounds may remain that upon accumulation in natural waters can be harmful as well as a nuisance. At the time of this writing, mercury has succeeded phosphate as the major aquatic villain. What element will be next? It hardly makes a difference. The present danger with fish kills from mercury and overfertilization from phosphate strongly suggests that secondary treatment is insufficient. What we really need is tertiary treatment—or recycling of our waste water—so that the effluent (discharge) from our waste water plant is literally good enough to drink. This will involve both chemical and biological effort beyond the present treatment. Let me mention one novel approach with which I have had some prior contact.

Biological Phosphate Removal. In 1966 Gilbert Levin published a method for utilizing activated sludge to remove phosphates (or to lower it to acceptable levels) from waste water. He based the removal on a concept he called "luxury uptake" in which microbes in sludge, when aerated very vigorously, took in more phosphate than they needed for growth, ergo "luxury." These phosphate-glutted cells were then treated like sponges, the phosphates being "squeezed" out, so that the cells could be used again in another cycle (Fig. 45). In practice this procedure has not been uniformly successful, but it does demonstrate what is possible when ingenuity and microbes get together. We have now disposed of the treated effluent and it is as pure as we can get it, but what about all the accumulated sludge? The city of Chicago, for example, produces almost one thousand tons of dry solid waste daily. Burning is no answer—remember air pollution!

From 1968 to 1970, the city of Chicago waste treatment facility ran an experimental six-acre site in central Illinois for evaluating digested sludge as fertilizer. Under such treatment, corn yields have been seven metric tons per hectare, the equivalent of chemical fertilizer (Fig. 46). Certainly, this approach is not entirely innovative and is not without its drawbacks, but modern technology is making it possible to reexamine some ideas previously discarded due to technical problems.

RECYCLING

Do you recall the line from a popular Cole Porter song, vintage

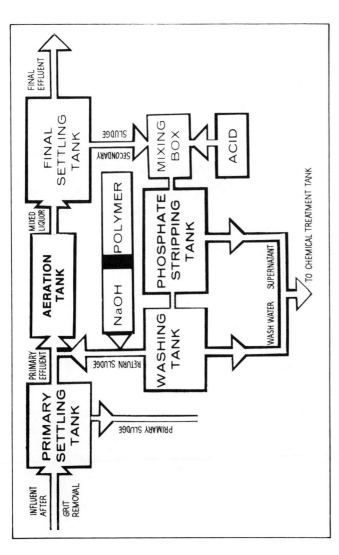

45. Flow diagram of pilot plant for Biological Phosphate Removal
In the aeration tank the microorganisms in the form of activated sludge take up phosphate. The excess phosphate (luxury uptake) is removed in the phosphate stripping tank by the addition of acid. This phosphate can then be treated separately. The sludge is returned for reuse.

46. Sludge Being Thrown on Corn Fields
Reproduced, by permission, from New York *Times,* May 28, 1972.

World War II, "Give me land, lots of land, under starry skies above. Don't fence me in"? If cattle could sing, this might be their dirge. No more do longhorns roam the range until market time. We do not have the range space, nor the time it takes for fattening on the range: nor do we especially like the flavor of range-fattened beef. Beef cattle are now being processed to a greater and greater extent in feed lots, containing as many as 100,000 head with as little as 50 square feet each for roaming space. The name of the game is weight gain: 50 pounds in three weeks. This rapid weight gain precludes rumination as a major contribution to feedlot cattle nutrition. Their confinement and high protein diet results in heavy manure density, very rich in unutilized nutrients. The estimates for annual total feed lot manure in this counry hover at the one billion ton figure. Its disposal is indeed a big problem.

Two proposals, both relating to microbial productiveness, are being examined seriously. In addition to undigested, still usable feed, 30 to 40 percent of manure consists of good protein-containing microbes. The undigestable lignin and cellulose are being separated from the manure. The remainder then is processed for re-feeding. The other feasible approach involves anaerobic digestion. Methane gas in economically recoverable levels has been produced in six days; one third of the organic load was reduced to 50 percent methane.

Mechanization has also hit the poultry industry. Millions of fryers and laying hens never see the light of a real dawn. Instead, they spend their lives in chicken-wire prisons, containing tens of thousands of jail birds, all doing their thing— chicken manure (ten chickens= one man) is no small thing— and cannot be disposed of at twenty-five cents a bushel as in the good old days. Below ground level, immediately adjacent to or underneath the chicken factory, many processors have constructed bacterial oxidation ditches where the organic content is reduced to safe levels, much as I described earlier in the activated sludge process. Another method of disposal involves liquid soil injection. A manure slurry is pumped into furrows and is conveniently covered with soil. This prevents insect and odor problems and, in addition, provides excellent fertilizer for conversion by soil microbes.

I have discussed with you the problem of disposal of waste from large populations of animals raised or fattened together for economical purposes. There is a contrasting situation in which waste from isolated animals is collected and brought to one place advertently. The best substrate (in this case, organic soil) for commercially raising *Agaricus campestris* (the edible mushroom) is based on composed manure. What is compost? Imagine manure from 30 to 40 riding academies and 3 race tracks, and the urine-soaked bedding from the stables, piled in a mound 100 by 200 by 10 feet. The indigenous microbes begin attacking the organic material, particularly at the oxygen-free center.[4] Temperatures as high as 70°C are developed and the original components begin to lose form. A small bulldozer (or in the old days, many pitchforks) turns the composting mixture at daily intervals so that the entire compost pile becomes digested. After a week to ten days, digested compost takes on the appearance of humus. It is then commercially steam-sterilized to kill its microbial population, placed in flats in dark, high-humidity rooms, and seeded with mushroom spores. White mushrooms sprout forth, adding their nutrition and delectable taste to our already crowded microbial menu.

I close this chapter with a few words on refuse (garbage and rubbish). It has been estimated by the Department of Health, Education and Welfare that the average American deposits 3½ pounds of refuse daily, of about 400,000 tons daily for the nation, a mountainous amount. There are two general methods for disposing of refuse that involve the activity of microorganisms. The first, sanitary landfill, I am sure is well known. Essentially, a big hole is made in the ground and the refuse is dumped and then covered. The degradation is just about what I described earlier in the chapter. The other method involves composting, in which the refuse is ground to a uniform size, piled to a reasonable height to allow anaerobic activity internally, and turned regularly to complete stabilization. This method has been widely adopted in Europe and is being more seriously investigated in this country at the present time. A novel method of increasing composting activity combines animal waste with refuse. I am sure this approach is possible and feasible. Whether it gains widespread acceptance remains to be seen.

Regardless of whether microbes are utilized or not in any specific degradation process, in the final analysis it is the microbe that will make the ultimate conversion. As Louis Pasteur once said, "Messieurs, c'est les microbes qui auront le dernier mot!" ("Gentlemen, it is the microbes who will have the last word!").

EPILOGUE
FIRST IN WAR—
FIRST IN PEACE

I have reserved discussion of infectious diseases outside of the mainstream of this book because calling a pathogenic microorganism a friend, unseen or not, might require explanation. Throughout history nations have waged war against each other. Historians have chronicled many wars and battles in which disease played a decisive role in determining the outcome. Thus, to the winner (when there was one) the culprit microbe was, indeed, a friend. I have also decided to play the devil's advocate for biological warfare, an aspect of which is currently gaining some attention. Please defer judgment on this subject until later in the chapter.

For better or for worse, microbes have been a determinant in establishing certain mores in art, literature and behavior. Perhaps with tongue in cheek, you may be able to accept the role of microorganisms as beneficial, even though the short-term outcome of their activities may have been far from friendly.

The first instance in which disease played a decisive role in the outcome of a conflict, I have selected from the Bible. In the Louvre in Paris there is a painting, the *Plague of the Philistines,* in which the careful observer can see little rats scurrying about. The implication is that this plague is *The Plague,* that is, the one caused by the bacterium *Yersinia pestis* and carried by the rat flea, *Xenopsylla cheopis.* Samuel 1:5 describes in detail what happened when the Philistines, apparent victors in their fight with the Hebrews, made the mistake of kidnapping the Ark of the Covenant.[1] "The wrath of the Lord was against the city with a very great destruction, and He smote the men of the city, both small and great and they had *emerods* in their secret parts and the hand of God was very heavy there, and the men that died not were smitten with the emerods." "Emerods" has been variously translated as "hemorrhoids," which could be considered in the secret parts, but it is difficult to conceive of this discomfiture causing such a high mortality rate. Another meaning is swelling or protuberance, a more logical assumption since the inguinal lymph nodes do swell in the plague; such swellings in the groin are referred to as buboes (Greek *boubon,* "groin"), thus the name bubonic plague. So, the first savior in the written tradition of the Hebrews was one of our unseen friends (no irreverence intended).

One of our most able chroniclers of the past was the Greek historian Thucydides, who among others has so adequately described diseases that modern medical historians have been able to give them scientific labels. Their descriptions were not always foolproof, however, since symptoms of many infectious diseases have changed through the centuries. In the second book of Thucydides it is mentioned that in the Peloponnesian wars in 430 B.C., the main reason that the Athenians did not attempt to expel the Lacedaemonians was due to the widespread debility caused by plague, typhus, and smallpox. Not only did Athens fall because of disease, the invaders themselves were forced to leave for the same reason. Disease killed 10,000 freemen and 45,000 citizens, including the famous Pericles.

In 296 B.C., the Carthaginians invaded the island of Sicily and laid siege to the city of Syracuse. The invaders were almost successful, but they were overcome by a smallpox epidemic. They were forced to leave this island and although this debacle preceded the Punic Wars by a century, its outcome might have been different had the Carthaginians been able to establish a beachhead in Sicily. According to the eighth book of Herodotus, in 400 B.C. the Persians under Xerxes, almost victorious over the Greeks, were forced to retreat to their native Persia because of an epidemic of dysentery which caused a 50 percent mortality among the Persian troops.

Even the victories of Marius over Octavius in the Roman Civil War in 88 B.C. could be attributed to the loss of 17,000 men by Octavius from smallpox. The famed Marcus Aurelius lost both his life and Gaul around 200 A.D. after his troops were decimated by disease. In the fifth century, the Huns were forced to slow their conquest outside of Constantinople because of plague and other diseases among the troops.[5] There is very little to tell about the details of history during the silence of the Middle Ages. In 1285, in the war of Spain against France, the Prince of Aragon, the Spanish leader, was killed by the plague. During the Spanish civil wars of the fifteenth century 20,000 deaths occurred in the army of Don Fernando, the Catholic, of which 17,000 died of typhus fever.

Perhaps one of the most dramatic examples of the impact of disease on a military incursion occurred in 1439. On October 1 of that year, Albrecht, the German emperor, had reached the walls of Baghdad. Two weeks later he was dead of dysentery and his army was in full retreat.

THE GREAT POX

1495 is an historic year; the army of Charles VII of France, while besieging Naples, was struck by an epidemic of syphilis. This was the first mention and appearance of this disease in the western world. The disease was also widespread among camp followers and Neapolitans. At this point in history the etiology of this malady was unknown and its effects were demoralizing on the troops of Charles VII. It has never been definitely established whether the citizens of Naples were the transmitters or the receivers of that disease. No one wanted to take credit for it. The Italians called it *Morbus Gallicus,* the French disease, obviously blaming it on the soldiers of Charles VII; not to be outdone, the French referred to it as *la maladie Italienne,* the Italian disease. The English were neutral, they called it the Continental disease. Coincidentally, its explosive appearance in epidemic proportions came just three years after the first voyage of Columbus, whose men and/or the aborigines of the West Indies were for many convenient scapegoats. Although the mystery of its ultimate origin still remains, the method of transmission was lyrically detailed by the Italian natural philosopher Giralomo Fracastoro (Hieronymous Fracastorus) who in 1530 wrote the epic poem, *Syphilis Sive Morbus Gallicus* ("or the French Disease").[6] In iambic hexameter, he describes how an innocent shepherd named Syphilus was cursed for angering the gods with the disease that now bears his name. Fracastoro described in poetic detail its epidemiology and symptomology for all time. His chef d'oeuvre has been translated into five languages and has been published in some thirty editions. Thus the chance philanderings of an unsuspecting adolescent who momentarily left his pastoral responsiblities has given him unrecognized immortality. In the sixteenth century, the disease became an epidemic scourge to which its pseudonyms give mute testimony; *Lues Venerea* or *Lues,* the Venereal plague, or the pox, literally relegating the diminutive adjective to the more famous *small pox.* Pray tell, whose friend is *Treponema pallidum,* the spirochete causing syphilis?

Let us examine the evidence. Ever since the army of Charles VII was unawaredly defeated by this slender microscopic love bug, armies have been consistently plagued by its incapacitating effects. If the adage "An army travels on its stomach" has a ring of truth, one must be tempted to add "also on its back." It has been suggested and not without a note of seriousness, that purposefully infecting an opposing army with venereal disease might be a convenient method for achieving a bloodless victory. Certainly, for a long time the military took a punitive view of this kind of infection. Until 1945 the U.S. Army deprived soldiers of pay and possibly rank during the period of their illness. I had ample time to observe the results of this rather primitive policy. In May 1945 I was the bacteriologist in a field hospital attached to the First Army and then the Fifteenth Army near Bonn, Germany. When the war ended I spent practically my entire working time doing laboratory diagnoses for syphilis, and its bedfellow, gonorrhea,[7] and helping the soldier-victims fill out epidemiology report forms listing the who, where, and when of their encounters (an aside: no fraternization with the German civilian population was allowed and more rendezvous were consummated on the roads between Bonn and Cologne with unknown Russian and Polish displaced persons than could be recounted). These poor men were more concerned and terrified about what the military would do to them than what their medical prognosis was. Shortly thereafter, the policy towards venereal disease was modified with the condoning and sanction by the military of "illicit sex." If a soldier had a pass stamped "prophylaxis taken" at an official station off base at the appropriate time preceding the appearance of the disease, he was excepted from punishment.

I have no doubt that in the last twenty-five years this rather incongruous position has been again modified and substituted by a saner, more humane one. Notwithstanding, venereal disease is still the bane of the military. We can thank syphilis for several landmarks: the poem by Fracastoro, the first scientific treatise on the epidemiology of a disease, the invention of the condom[8] (the first preventive measure for a specific disease), the development of the Wasserman Complement Fixation Test, the first "blood" test for a specific disease, and the discovery of Salvarsan by Paul Ehrlich, the first chemotherapeutic agent for a specific disease. Indirectly, both the condom and the Wasserman test have helped forestall a related social problem. The Wasserman test was the forerunner of one of a number of premarital blood tests specifically required by law in most "civilized" parts of the world. The delay imposed by this test may be responsible for preventing many couples from precipitously plunging into matrimony. After many years of practical if not esthetic success as a birth control device, the condom has fallen on hard times. The rise of the contraceptive pill has unshackled sex and relegated the condom to its original use, for which it is more rarely used. Thus in the past several years, as the woman has been liberated, the man has simultaneously dropped his defenses. So, while the birthrate has not suffered an increase, venereal disease rates in our times are setting records (14 million cases in 1969 and rising). In 1975 we have come full circle: rising VD rates and potential dangers of the pill have brought new popularity to the condom.

OFF TO THE WARS

The famous Thirty Years War and the struggle between Charles I and the Earl of Essex, both in the seventeenth century, were brought to conclusion by disease among the ranks of the vanquished. One hundred years later, the Swedes had successfully penetrated Russia and the French had reached the gates of Prague. Both invaders were forced to retreat ignominiously because of plague and smallpox. Other outstanding examples of military misadventure caused by communicable disease are the several disasters of Napoleon Bonaparte. From his ill-fated attempt to put down the colonial uprising in Haiti to his infamous retreat from Moscow, he was literally plagued. The results of that 1812 venture were a reduction in his army from 500,000 to 20,000 able-bodied men; most of the casualties could be attributable to dysentery, typhoid, and typhus, rather than enemy bullets.

I think by now you get my message. There will be more wars and where there is war there is that other apocalyptic horseman, disease; and although for the short run there appears to be a winner, wars do not have winners, only losers.

This history has not been lost on certain defense-minded individuals with foresight and perspicacity. Why leave the time, place, nature and victim of a disease to chance? And thus was born biological warfare; just one more example of man's inhumanity to man. I am pleased to say that at this writing, the military direction of such an establishment in this country has been halted even though a great deal of very valuable basic research has been generated in its name in the prevention and detection of communicable disease.

The possibility of microbial intervention in wars has been suggested for several instances of which the stories have been retold so often that the retellings have achieved legend status. Two deserve repetition here.

During the American Civil War (1861-65) the ascendancy of Ulysses S. Grant as a field general of the Union army was marked with bitterness, controversy, and considerable envy from contempory fellow officers. Grant's career was somewhat erratic following his graduation from West Point due, in part, to a drinking problem. However, with the outbreak of hostilities and the reassumption of an active command, Grant made a vow to avoid alcohol in all forms. In fact, he gave his word of honor to a close friend, Assistant Adjutant General John A. Rawlins, that he would touch not a drop for the duration. He made one slip according to Rawlins, during the Vicksburg campaign, which he regretted and which was without consequence in the military sense. Notwithstanding, stories of his drinking were legion and were viciously circulated by his enemies to discredit him. Such rumors led to the famous interview given by President Lincoln which assumed that not only was Grant drinking heavily and that Lincoln condoned it, but that this behavior was responsible for the military successes.

After having heard from many reliable sources close to Grant that tales of his drinking were unfounded, Lincoln made a plan to stop the rumors. Carl Sandburg quotes Lincoln as having said,

One day a delegation headed by a distinguished doctor of divinity from New York, called on me and made the familiar complaint and protest against Grant being retained in his command. After the clergyman had concluded his remarks, I asked if any others desired to add anything to what had already been said. They replied that they did not. Then looking as serious as I could, I said, 'Doctor, can you tell me where General Grant gets his liquor?' The doctor seemed quite nonplussed, but replied that he could not. I then said to him 'I am very sorry, for if you could tell me I would direct the Chief Quartermaster of the army to lay in a large stock of the same kind of liquor and would also direct him to furnish a supply to some of my other generals who have never yet won a victory.'[9]

It was obvious that Lincoln was being his humorous best. "What I want," he later said, "and what the people want, is generals who will fight battles and win victories. Grant has done this and I propose to stand by him. I permitted this incident [the above] to get into print and I have been troubled no more with delegations protesting against Grant."

MILK AND HONEY FROM ACETONE

A much better case can be made for microbial intervention in the following sequence. Dr. Chaim Weizmann was a Russian chemist who immigrated at the turn of the century to England, where he became a reader in chemistry at the University of Manchester and at the same time a part-time researcher at the Clayton Aniline Works. He was an active Zionist while pursuing his scientific career and discovered a strain of a bacterium, Clostridium acetobutylicum, which produced acetone and butyl alcohol from miscellaneous carbohydrate sources. This made acetone available during World War I as a solvent for naval gun explosives, thus contributing much to the Allied victory in that war. As a reward, he was given the Balfour Declaration which opened Palestine to world Jewry after World War I. In 1948 he became the first president of Israel.

The above sequence is approximately true. However, in his autobiography[10] Weizmann takes issue not with the events but with some of their inferences. In 1904 he did, indeed, begin a teaching and research career in Manchester and in 1906 he met David Balfour, who was standing for Parliament from that district. It was the beginning of a friendship that would, indeed, culminate in the famous Balfour Declaration of 1917. However, Weizmann's Zionist activities perhaps influenced Balfour more than his scientific contributions, for it was deemed in the best interests of British foreign policy and the War Cabinet's political future to support a Jewish homeland.

Let us return to microbiology. In connection with his teaching reponsibilities Weizmann became interested in the new science of bacteriology and made several visits to the Pasteur Institute before 1904, and in addition worked with the famous Swiss bacteriologist, Burri, on bacteria in milk. In 1910, while looking for a precursor for synthetic rubber in a fermentation reaction, he instead got a culture yielding a mixture of acetone and butanol. He put this discovery aside as being of little practical significance. At the end of 1914 the British War Office invited every scientist to report discoveries of military value. Weizmann offered his fermentation scheme but no one was interested. In that same year he met the future Prime Minister of Great Britain, Lloyd George, who although he does not acknowledge having met Weizmann until 1917, does suggest in his memoirs that the Balfour Declaration was indeed a reward to Weizmann. It was not until late 1916 that Weizmann's discovery of a method of acetone production came to the attention of the British government officially. The supply of acetone as a solvent for cordite was fast being depleted and the previous major supplier had been the I. G. Farbindustrie in Germany. Without acetone British naval guns would have needed to be redesigned. Weizmann was told by Winston Churchill, the First Lord of Admiralty,[11] "We need 30,000 tons of acetone." With the help of the brewing industry and Canadian and American industrial support this was accomplished. Several sources of carbohydrate were available but due to the vast shortage of food in the British Isles the best substrate for *Clostridium acetobutylicum* could not always be used. Weizmann even resorted to the use of horse chestnuts as a carbohydrate source, but in that case was only partially successful. The Royal Naval Cordite Factory in Holton-Heath, England was able to get the following yields from a thousand pounds of maize (650 pounds of starch): 163 pounds of butanol, 70 pounds of acetone, 407 pounds of CO_2, 11 pounds of hydrogen and 12 pounds of miscellaneous acids. After the war, Commercial Solvents Corporation bought the patent rights from the British government and Weizmann was giving a token reward of ten shillings per ton of acetone which amounted to a total of 10,000 pounds sterling (about 50,000 U.S. dollars).

During the entire war period Weizmann, who was president of the Zionist Organization of England, never ceased in his political efforts on behalf of a homeland for world Jewry. On November 2, 1917, Lord Balfour, who was now Foreign Secretary, issued the now famous Balfour Declaration: "His Majesty's government views with favor the establishment in Palestine of a national home for the Jewish people and will use their best endeavors to facilitate the achievement of this objective, it being clearly understood that nothing shall be done which may prejudice the civil and religious rights of the existing non-Jewish communities in Palestine or the rights and political status enjoyed by Jews in any other country." Notwithstanding Weizmann's refutation that the declaration was a reward to him for services rendered, I am sure that his scientific eminence, which brought him to the attention of prominent men, as well as his staunch advocacy of Zionism, played equal roles in that very important piece of history. How important could not be measured at that time nor even in 1948 when statehood for Israel was declared and Dr. Chaim Weizmann was named its first president. Today, fifty-nine years after *Clostridium aceto-*

butylicum produced its vital acetone and Balfour produced his controversial declaration, the drama is still unfolding on center stage in the Middle East.

MODES, MANNERS AND MICROBES

As we have seen the tide of battles influenced by the permissive activities of some microbes, so we can also view how some have affected style and thought, art and literature.

During the year 250 A.D., mass conversions to Christianity were brought about by St. Cyprian during a plague epidemic, conspicuous because of the wearing of black as a mourning symbol for the first time in the western world. Could it have been a warning to keep one's distance, signifying that a relative had suffered the Black Death? The plague inspired many writers to chronicle their reactions to the horror and panic devastating their times.[12] The *Diary* of Samuel Pepys starkly reports (London, June 7, 1665): "People left London in innumerable multitudes. Servants, shopmen, clerks, turned out left homeless and helpless, wandered around without habitation." Surrounded by burying carts, unburied bodies and open graves, Pepys commented, "The hottest day I've ever felt in my life. I was put in a poor conception of myself and my smell so that I was forced to buy some roll tobacco to smell and chaw which took away my apprehension." The population of England and Continental Europe was decimated and terrorized; this plague was not forgotten for some time. Daniel Defoe (of *Robinson Crusoe* fame) in 1721, recreated the whole scene in *The Journal of the Plague Year*. Perhaps in a lighter vein is the *Decameron* (circa 1350), by Giovanni Boccaccio, a fictionalized recapitulation of tales, sometimes bawdy, told by ten young people (seven women and three men) on the outskirts of Florence, Italy, during another plague epidemic. They spent ten days together moving from one country place to another regaling each other with delightful tales while they waited out the plague.

Our medieval ancestors were much taken with the

most drastic of the infectious diseases since they associated those so afflicted as cursed. Certainly leprosy, which by this generation has greatly wanted in intensity, attracted a great deal of attention. Lepers were shunned and segregated. They lived and begged together and there was even a Lepers' Guild. In the Louvre (if you are lucky), you can see Botticelli's *The Offering of a Leper.* Look closely and you will notice the missing fingers, one of the prominent disfigurements of this disease, now known to be caused by *Mycobacterium leprae.* Caring for the leper was an extreme act of contrition; witness the painting by Hans Holbein *Saint Elizabeth Washing the Leper.* Even Rembrandt Van Rijn had his *Leper.*

If Bubonic was the *Black Plague,* then surely tuberculosis earned the title, the *White Plague.* This disease, shown by Robert Koch in his classical demonstration of the germ theory of disease to be caused by *Mycobacterium tuberculosis,* has had many faces through history. In the Middle Ages, scrofula (tuberculosis of the cervical lymph nodes) was prevalent. The adjective *scrofulous* was used not only to describe the disease but also to describe a morally contaminated individual. It was also known as the "King's evil," perhaps because even royalty were not immune.

The disease ran many courses from the tenth to the twentieth centuries. The King's evil was first "curable by the King's touch." Even Shakespeare got into the act.[13]

The scars from cervical lesions were hidden by the fashions of the times, as the high-yoked dresses and blouses of the fifteenth to the seventeenth centuries testify (Fig. 47). Of several well known paintings, I will just mention that of Franz Hals, *The Portrait of a Woman,* and the painting by Peter Paul Rubens of his brother. Although many forms of tuberculosis existed side by side, the dramatic shift from cervical to pulmonary tuberculosis occurred coincidentally with the onset of the so-called Grecian revival period around 1800. The romanticists had arrived! Compare Franz Hals's woman with *"Pinky,"* by Sir

47. *Portrait of a Woman*, by Franz Hals
Reproduced, by permission, from Detroit Institute of Arts, Michigan.

Thomas Lawrence (Fig. 48). "Pinky," delicately poised, the moors in the background, a gentle wind wafting her diaphanous gown, is a sure case for respiratory infection. Despite the apparent gaiety of this immodest miss, the era was obsessed with morbidity. To call it Romantic seems incongruous. Its most notable poets were incessantly writing of death and quite coincidentally most of them had the consumption (pulmonary tuberculosis). Keats died from it, Shelley suffered from it, and Byron longed for it.[14] To this famous trio add Robert Louis Stevenson, Ralph Waldo Emerson, Henry Thoreau, Robert and Elizabeth Browning, The Brontë family, Alexander Dumas, Charles Dickens, Edgar Allan Poe, and others ad infinitum. The heroine of Dumas's La Dame aux camélias (Camille), Marguérite Gautier, was a consumptive who was further immortalized by Verdi as Violetta in La Traviata; Puccini did likewise for poor Mimi in La Bohème. Although no professed consumptive, Bunthorne, the hero of the operetta Patience, epitomized the pale, wan lover of the period and was the Beau Geste of Gilbert and Sullivan.

To be consumptive was to be wanted and loved. Cosmetics and diets to mimic the physical face of the disease were common. Such genius was associated with its ravages that a theory was even proposed that the tuberculosis itself was responsible for increasing intellectual capacity. Notwithstanding the fact that to the list of literary giants we can add the likes of Chopin and Paganini, such a theory falls far short of being proven. Perhaps what can be said is that talented and gifted people were spurred on to prodigious feats knowing that their affliction had condemned them to a premature death.

With the increase in the mortality rates from tuberculosis and with the explosion into the bleakness of the industrial era, the glamour of consumption wore off and its true nature as the killer of the poor and a destroyer of all social classes became more and more evident.

Let me leave my flights of fancy into fact and fiction and end this chapter with a discussion of a serious and valid

48. *Pinky*, by Sir Thomas Lawrence
Reproduced, by permission, from the Huntington Library and Art
Gallery, California.

contribution of microbes to the solution of a current problem. I mentioned earlier that I would defend biological warfare at the appropriate time; that time is now. The following headline appeared in the *Washington Post* August 26, 1970: USDA PLANS TO WAGE GERM WARFARE AGAINST COTTON BOLL-WORMS IN SOUTH. We have come full circle. The father of modern microbiology, Louis Pasteur, was called upon in 1865 by the sericulturists (silk-worm raisers) of France to solve the mystery of *pébrine*, the disease that was killing the silkworm. Disaster threatened. Louis rose to another occasion; these industrious caterpillars would once again eat their way "round the mulberry bushes." Diseases of insects are two-edged swords. When the insect is a valuable commodity like the honey bee or the silkworm, prevention of disease is the order of the day; but where the insect is a commerical or a public health nuisance, its diseases, especially if fatal, are a blessing. Today the use of selected microorganisms against specific target insects is a valid alternative to the broader spectrum persistent chemical insecticides like DDT.

In 1915 a German entomologist, Berliner, isolated a bacterium from dead and dying larvae (caterpillars) of the storage grain moth (*Ephestia kuhniella*). He determined that this microbe was the cause of their deaths and named his discovery *Bacillus thuringiensis,* in honor of the province of Thuringia in Germany. During the twenties and thirties *Bacillus thuringiensis* was used to a limited extent for experimental control of the European corn borer. In 1951, E. A. Steinhaus rekindled interest in microbial control of insects. There have been other notable landmarks. In 1941, Sam Dutky, of the U. S. Department of Agriculture, introduced *Bacillus popilliae,* the cause of milky disease of Japanese beetle larvae, as a control measure for that insect. When applied it has been extremely successful. Some fungi have been investigated, but with mixed results. Present enthusiasm centers on specific insect viruses (re: headline from *Washington Post* in opening sentence of this section).

THE DIMENSIONS OF THE PROBLEM
Before I detail the specific microbial insecticides and how they would function, it would be appropriate to state the dimensions of the problem being attacked, and the advantages and limitations of microbial control.

Insects are the most numerous of all animals; at least one million species are known. No more than 15,000 are considered nuisances and of these, 400 at most are destructive or dangerous enough to warrant serious control measures. Two distinctive groups are involved: insects destroying crops and forest lands, and those of public health importance, whether directly or as vectors of disease. I will direct my attention to the first of these two.

By their very nature, insect pathogens used as insecticides have a much narrower range of activity than chemicals, thus not only are they less likely to create difficulties for higher animals, they also tend to discriminate even among insects. Thus, it is possible to aim the microbial insecticide at a specific target insect. However, the ability to kill an insect is only the beginning. There are several corequisites that are equally as important as lethality:

(1) Virulence should be of a high order and not readily lost.

(2) Time between infection and cessation of feeding should be sufficiently short to preclude damage to plants.

(3) The insect pathogen should be harmless for man, other vertebrates, and beneficial invertebrate animals.

(4) It should have physical and chemical stability during storage.

(5) It should be economically feasible to produce.

How far have we come toward meeting those criteria with specific control organisms?

The outstanding example of successful biological insect control is the use of *Bacillus popilliae* for the Japanese beetle. Where it has been used adequately, the beetle has disappeared. The larvae burrow in the ground, become infected and as the bacillus multiplies in its body cavity, milky disease develops. Death follows with the release of literally millions of bacteria. This process is repeated many times so that the top layer of soil becomes saturated with infectious microbes. There is one problem with *Bacillus popilliae*. It is a spore-forming species whose endurance in the environment is completely dependent on the formation of that spore. Sad to say, although much research has been and is still being done, *Bacillus popilliae* does not produce either mature spores or adequately infectious cells when grown in culture medium. Only when the beetle larvae themselves are used have positive results been achieved. Due to its residual nature, this has not proven a drawback to control. In other ways this means it meets all the criteria previously mentioned.

THE DIAMOND IN THE ROUGH

During the last twenty-five years, strains of *Bacillus thuringiensis* have been isolated all over the world in Siberia, France, the United States, Japan, Czechoslovakia and Canada. Susceptible insect species all belong to the order Lepidoptera (moths and butterflies). In 1953 Hannay, and independently, Angus, unveiled the mystery of the microbial lethality (Fig. 49); within each bacterial cell in juxtaposition to a heat resistant endospore resides a diamond-shaped protein crystal. Angus and Heimpel further found that this crystal dissolved in insect stomach juices (more precisely the mid-gut at a highly alkaline pH) causing a paralysis that eventually led to its death. Subsequently, well over one hundred caterpillar species have succumbed to varieties of these so-called crystalliferous bacilli (see Appendix).

Although the primary mode of action against caterpillars is due to the crystalline endotoxin (*endo*, "inside the cell"), several strains of crystalliferous bacilli also produce an exotoxin (*exo*, "outside of the cell," i.e., excreted as a waste product). In contrast to the protein crystalline endotoxin, which is heat sen-

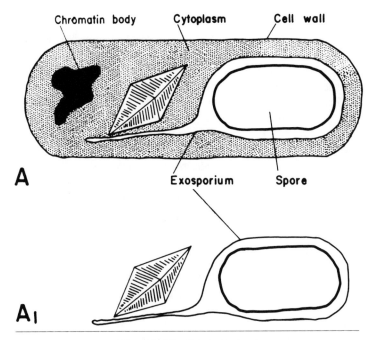

Chromatin body   **Cytoplasm**   **Cell wall**

A

Exosporium   **Spore**

A₁

49. Spores and Crystals of *Bacillus thuringiensis*
(a) A diagram illustrating the position of the protein crystal relative to other structures during sporulation. (a₁) After completion of sporulation.
Reprinted, by permission, from Arthur Heimpel and Thomas Angus, "Diseases Caused by Sporeforming Bacteria," *Insect Pathology,* ed. Edward Steinhaus (New York: Academic Press, 1963), p. 36.

sitive (destroyed above 60°C), the exotoxin is nonprotein, heat-resistant, and effective against muscoid (related to the house fly) larvae. By feeding exotoxin-producing strains of *Bacillus thuringiensis* to chicken and cattle, some success has been achieved in controlling flies in their feces. However, the major insect hosts for *Bacillus thuringiensis* are still caterpillars and despite some of its drawbacks, there is sufficient interest in its development to have produced at least eight patentable products in four countries (see Appendix). Permits have been received from the

U. S. Department of Agriculture for use on a number of edible crops (see Appendix) and numerous products based on *Bacillus thuringiensis* have been submitted to the U. S. Environmental Protection Agency for possible registration. Field studies throughout the world have indicated that *Bacillus thuringiensis* has the potential to replace traditional insecticides in many areas. I had the privilege of participating in one such program for the control of gypsy moth larvae in Spain. In Figure 50 and Figure 51 you can readily see the protection afforded an important crop, the evergreen oak (the acorns are a major source of animal food), by treatment with a spore-crystal suspension of *Bacillus thuringiensis.*

FRIENDLY VIRUSES

The smallest of our unseen friends, the viruses, are being successfully used to control certain caterpillar species. Viruses that attack insects have been known for some time: however, using the basic information for insect control presents certain technical problems. These relate to the necessity of using a living system (other caterpillars or tissue-cultured from caterpillars) to grow them. Three types of viral infections are described for insects and although all have probable use, the nuclear polyhedrosis viruses have been exploited to the greatest extent. In nucleopolyhedrosis infection, the virus multiplies in the cell nuclei of the larvae which then become swollen with polyhedral bodies.[15] The cells eventually break open, releasing the polyhedra, which can be seen in the ordinary light microscope. The dying larvae climb to the highest point they can reach and characteristically hang head-down until dead. As of 1970 almost 300 viruses with biological control potential had been isolated, 18 of which have been sufficiently tested to recommend their immediate use. One of these I referred to in the beginning of this section: the Food and Drug Administration has issued a two-year permit for the use of a nuclear polyhedrosis

50. Untreated Branch of Evergreen Oak Showing the Results of Gypsy Moth Larvae Feeding

51. A Branch of Evergreen Oak Sprayed with Suspension of Spores and Crystals of *Bacillus thuringiensis*.
The foliage is intact.

virus against *Heliothis zea* (known to cotton farmers as the cotton bollworm, and to corn growers as the corn earworm).

Thus, germ warfare is really with us. However, much research remains to be done. We know too little of alternative controls for insect vectors of disease. Methods of propagation in large scale of all potential control agents are needed, as are more reliable and specific means of dissemination. More research must be done if this important area is to be fully developed. This country has a most admirable establishment available in which a great deal of the necessary research on biological control of insects can be done. On November 25, 1969, President Nixon totally renounced offensive biological warfare. This sounded the death knell for Fort Detrick, Maryland, as a continuing biological warfare center. It is the most complex and complete facility for isolation, containment, and mass rearing of pathogens to be found anywhere in the world and is located near the major research station of the Department of Agriculture. How fitting if this magnificent establishment could be turned from one kind of biological warfare to another, one that would benefit mankind! Perhaps this change in direction could be the first of many that might help fulfill the prophecy of Isaiah (2:4): "They shall beat their swords into plowshares and their spears into pruning hooks; nation shall not lift up sword against nation, neither shall they learn war anymore."

THE MICROBE'S CONTRIBUTION TO BIOLOGY[16]

I have attempted in these chapters to expose you painlessly to the science of microbiology. Hopefully, this has been achieved with a minimum of trauma. An understanding of processes and activities has been possible without becoming enmeshed in too many technical details. Perhaps the least acknowledged but in the long run the most substantial contribution of microorganisms are the scientific concepts discovered with them. Their small size, rapid multiplication, and ease of handling has made them excellent research tools, and they have played a central role in a large number of basic studies in biology and medicine. These include the unraveling of the mysteries of hereditary transfer, which has been worked out primarily in bacteria, molds, and viruses and has resulted in the awarding of some dozen Nobel prizes (see Appendix), of how vitamins work in metabolism, and of how energy is transferred in the living cell.

The microbe has been the unwitting partner of the biologist in other meaningful contributions, not least of which involves the etiology and cures for cancer. A current popular theory on the viral origins of cancers, in part, is based upon the genetic latency concepts developed from the relationship between bacteria and bacteriophage. The highly fermentative activity of tumor cells was first noted by the famed Otto Warburg (a previous winner of the Nobel Prize), who compared them to genetically deficient yeast cells that behaved similarly, asserting that perhaps the origins of malignancies arise in part from a loss of respiratory ability (loss of the Pasteur effect mentioned in Chapter 1). Many of the most successful anticancer chemicals, known as antimetabolites, received their preliminary clearance after successfully inhibiting the growth of selected microbes.

The microbe has proved its respectability and is entitled to a closing accolade: "The role of the infinitely small in nature is infinitely great" (Louis Pasteur).

APPENDIX

Prologue

Ministry of Technology, Torry Research Station, *The National Collection of Industrial Bacteria, Catalogue of Strains,* 2nd ed. (Edinburgh: Her Majesty's Stationery Office), 1966, is a collection of organisms of primary concern in industry in the area of deterioration and manufacturing. The species listed in this catalogue are undoubtedly the most closely related to the activities described in the book.

Chapter 1

Although I have by no means discussed all processes for wine in Chapter I, I believe you probably have a sufficient overview to appreciate what it takes to make a fine wine.

There is also considerable interest in the origins of wine, and I thought it might be worthwhile to include in the Appendix some selected background on a few of my personal favorites.

France: France has several wine regions excelled by no other areas in the world for the vintages they produce. First among these is Bordeaux, whose name actually refers to the Gironde Département of which Bordeaux is the capital. Labeling of wines, especially here, is a zealously guarded privilege. Wines are named for districts whose boundaries are rigidly enforced. These districts are granted *appellation contrôlée* (registered name). For example, Sauternes means only from Sauternes district; Médoc, only from the Médoc district. These bottles bear the label of the shipper or wholesaler.

In addition, the best of the regional wines most coveted are not blends from within the region; they come from one holding or château. Here, the grape is harvested, fermented, and bottled, and in grand vintage years they are the superlatives. The label *Mise en bouteille au Château Mouton-Rothschild* means that from the beginning to the end, this is the best there is.

The wines of Bordeaux are of two types—red and white—and primarily are made with the cabernet and semillon varieties respectively. Space does not permit a detailed listing and description of all types and labels, but I would like to mention the most prominent. Among the reds: Médoc, Margaux, St. Emillion, and St. Julien stand out, while the white wines of quality are primarily from Sauternes and Barsac. Most of the great château wines of general regions come from these six communes.

Not all the wines of the Bordeaux are privileged to bear even the Commune label. Second and third pressings and pooled yields in poor years qualify as *vin ordinaire* and may be sold as *Bordeaux Supérieur, Vin Blanc,* or *Vin Rouge*. Some of it is quite acceptable, especially if consumed from a carafe in a café on the banks of the Gironde. However, the majority of that bottled for shipment to the U.S. is not worth the effort.

Burgundy: This region comprises what was once the medieval kingdom of Burgundy and now covers three Départements in east central France. Perhaps the red wines of Burgundy are the only ones that approach (and, to some, equal) the excellence of Bordeaux reds; however, the white wines, though good, do not compare with such as the great vintages of Sauternes. Red Burgundy tends to be a more robust, hearty wine, perhaps not as delicate as its Bordeaux counterpart. The white wines are not as fruity and are a little drier. Burgundy suffers more from the vagaries of weather, and in a poor year the grapes are deficient in sugar. Thus the practice of chaptalisation (sugar addition) is sometimes necessary.

The best of the red wine is made from the Pinot Noir grape and the white from the Pinot Chardonnay. Because of vintage variability and the lack of large vineyards, wines of this region are usually sold under a district designation, in which growers pool their wines and sell to a reliable shipper. Among the more well known *appelations contrôlées,* you will find Pommard, Maçon, Gevrey-Chambertin, and Beaujolais representing the reds and Chablis, Meursault, Montrachet, and Pouilly-Fuissé, the whites. For the novice in wine-imbibing, the simplest way to distinguish a Bordeaux from a Burgundy is by the bottle. Traditionally, Bordeaux wines are in straight bottles with an abrupt shoulder, while Burgundies are found in bottles wider at the bottom that gently taper to the neck.

Other wines of France: Wine grapes are grown in every corner of France, and it would be a labor of love to devote this book exclusively to that subject. Unfortunately I have a broader calling in this book, and am forced to be restrictive.

There are two other regions of distinction whose wines are more or less celebrated: The Rhône Valley or Côte du Rhône. The two most well-known breeds are Châteauneuf-du-Pape, a vigorous, full-bodied red wine with the highest miminum alcohol content (12.5 percent) of any *appellation contrôlée,* and Tavel Rosé. The former comes from the area around Avignon where five centuries ago the French antipopes resided; hence the rather romantic name: Château-neuf-du-Papes (Popes). Tavel Rosé is the yardstick by which all rosé wines are measured.

Loire Valley: These wines are the least known of those exported but not, I believe, because of quality. Too little is available. The still and sparkling wines of Touraine and Vouvray, from the right bank of the Loire are unpretentious, delicately bouqueted, which in good years (1955, 1964) rival the best whites of Bordeaux. Recently, Anjou Rosé from the old capital of the Plantagenet Empire and on the left bank of the Loire has begun to rival the better known Tavel if only in volume. This wine is relatively inexpensive and is an acceptable table wine.

Wines of Italy: Despite popular impressions to the contrary, Italy, not France, is the greatest wine-producing country in the world. Without the output of Algeria, continental France has to take second place, even if by a small margin, to Italy (together they account for more than 50 percent of the world's output). However, this is in volume only. Even the most ardent Italophile would concede that her wines take a back seat to those of France and Germany.

Most of Italy's wine is consumed domestically, and the overall climate, coupled with the generally hilly terrain, makes for a high yield of common grapes with little effort. Thus except for isolated pockets of distinction, most is only qualified as *vino di pasta*. I will list for you the better-known wines by geographical region.

Piedmont: The city of Turin (Torino) is the home of sweet vermouth. This is a product made from Moscato grape wine, and at least 70 percent of the finished aperitif legally must be natural wine. To this are added herbs such as wormwood, sugar, and brandy. It is interesting to note that Italian vermouth from Turin and French vermouth from Marseilles are both made near mountainous areas where herb-collecting is simplified.* Also from the Moscato grape comes Asti Spumanti, a rather sweet, sparkling wine. Although some of this wine is made by the *methode champagnoise,* most is made by the cheaper, faster *charmat* or bulk process in which the wine is carbonated and bottled under pressure. Last but not least, from Piedmont comes Barolo, a red wine made from the Nebbiolo grape. Aged at least three years in a cask, it improves even in the bottle and is ranked by experts with some of the best of French reds.

Veneto: Next to Chianti, the two red wines of this region, Bardolino and Valpolicello (both made from essentially the Corvina grape), are the best known outside the country. They are good solid representatives of their breed and are worth trying. Perhaps the best known of the white wines of Italy, Soave, comes from this area. It is somewhat reminiscent of Chablis and, indeed, is made in the French way: with skins of the white grape, the Garganega variety. All three of these wines bear a Veronese *appellation contrôlée,* offering the consumer an assurance of quality and place of origin.

Tuscany: This is the home of Chianti, a name that has been plagiarized too often. The popularity of Chianti is due to a number of factors but, most probably, to the large volume of wine allowed that label and to the attractive wicker-covered, round bottom. *Fiascos* are not only cheap but also a perfect container for preserving the freshness. True Chianti contains 60 to 70 percent Sangiovese must and is fermented according to a ritual devised by Barone Bettino Ricasoli, the first lord of Brolio Castle and still the source of the best Chianti available. To the wary drinker, look for *Chianti Classico,* made from a

* Simple for the Italians, maybe, but not for the French. The Director of Noilly-Prat Company in Marseille told me that herbs for French vermouth were not collected locally.

restricted hillside in Tuscany. Each neck is cordonned with a label bearing a rooster and a registry number.

Umbria: Only one wine is worth mentioning: Orvieto, white wine that resembles somewhat the Sauternes of the Gironde. It is easily recognized, bottled in wicker-wrapped *fiascos.*

Sicily: I cannot leave Italy without mentioning Marsala wine. This wine has the same general characteristics in use and many similarities in production as Port, Sherry, and Madeira, and shares still a third commonality—that of being started by the ubiquitous British.

Wines of Germany: All of the well-known wines of Germany come from the valleys of the Rhine and Moselle, and all are white wines, easily recognized by their tall slender bottles. The legal and traditional restricitons of German wine-making defy any reasonably short description. As a rule, the best of the wines are from the Riesling or Traminer grapes with the more common vintages from the Sylvaner variety. There are several restrictions on sugar additions, which accounts for the generally lower alcoholic count of the ordinary table wine.

A comment on generic names: Liebfraumilch, despite its charming name, is a pooled wine of no significance, while Mosselleblumchen is the most common of the wines of the Moselle. The Germans are very proud of their wines, and if you want better-than-average vintage, just look at the label. You will find the entire pedigree: the village where the grapes were grown, the name of the specific vineyard, the name of the grape, and the date when the wine was made.

Spain: Although Sherry is by far the most famous of its wines beyond its borders, Spain ranks third in world-wine production because of its less prestigious wines. It is a pity that so little of these others is available for export. However, if you are fortunate enough to go to Spain, look for the simple red wines of Castille—those in the north, the Rioja district, and those to the south of Madrid, Valdapeñas. Although the vintage wines of these areas are not comparable to the French, their full-bodied *vino común* (at twenty-five cents per liter) takes second place to none. Another wine of distinction is Montilla, from the region around Córdoba. This wine is made from the same grape and in the same manner as Sherry but is rarely fortified and is usually consumed at its natural alcohol content of 14 percent. There is no more delightful pastime than slowly drinking a cool copita of Montilla in an outdoor café on a palm-shaded sidewalk in Córdoba! Presently, small quantitites of Rioja are available in the U.S.A. It is relatively inexpensive. Try some, but avoid like the plague Spanish red wines originating from Tarragona. In the larger cities you may run across Montilla. If you are a Sherry drinker, chances are you will become an even more devoted aficionado of this Cordoban elixir.

Portugal: Port wine is synonymous with the western part of the Iberian peninsula. This is primarily a fortified dessert wine whose drinking was established as a cult by the British. It is made from the Rabigato grape for the white variety or the Tinto caõ grape for the more popular Ruby or Tawny Port, both of which are grown on the banks of the upper Duoro River. This is the region around the city of Oporto, the second largest city in the north of Portugal.

Good Port bears no relationship at all to most of the wines sold under that name. I was made a believer with one glass of vintage Port in the cellars of Sandeman and Company in Vila Nova de Gaia across the Duoro from Oporto. Although the table wines of Portugal are consumed in great volume by the local population, they have not achieved any prominence outside the country. However, recently the rosé wines from Portugal, particularly those with slight carbonation, have met with great acceptance in the English-speaking world. In fact, it is claimed that one label, Lancer's, is the largest single imported wine in the U.S.

Madeira: It is said that Madeira is the only wine that ever challenged Port's claim to be the Englishman's wine. If anything, it is more like Sherry with varieties from very bone-dry, Sercial; medium dry, Verdelho; medium, Bual; and the equivalent to the cream, Malmsey.

A unique treatment following fermentation gives this wine its distinction, perhaps. The young wine is stored in heated cellars (100°F). This encourages evaporation and aids in maturation. The sad truth of Madeira is that this small island cannot supply the demands for their royal vintage.

Wines from Here and There
Austria: The full-bodied, fruity white wines, excellent when

young, come from several areas in Austria. The richest yields come from that part of the Danube Basin known as the Wachau where one slope is called the 1000-Barrel Mountain. The most delightful of the vintages are Gumpoldskirschner and Grinzing, both near Vienna; the former is readily available in the U.S.A.

Hungary: Tokay wine from the furmint grape is another of those wines that bear no resemblance to their American counterparts. Noel Coward referred to Tokay in song as "Tokay, the golden sunshine of a summer's day." The wine is covetously restricted to a narrow region near the Czech border. Although traditionally a sweet dessert wine made from overripe grapes, the other varieties of Tokay, so labled, are available as medium and even dry forms.

The Good Old U.S.A.:

Last but not least in our tippling itinerary is the United States. Wine consumption in this country has been steadily increasing, as is the sophistication of the wine imbiber. Several states produce quantities of wine, but the best known are New York and California. The Finger Lake district of New York and the Napa Valley of California have a reasonable combination of soil and climate required to approximate some of the better wines of Europe

Sherry types are being labeled *Solera Process,* and the best of the domestic wines are varietal products; they are not called after a region of origin (for example, Sauternes) but after the variety of grape (at least 51 percent) used there (Semillon). This is an honest approach to labeling, since the grape is only one member of the triumvirate that includes weather and soil. Thus you can expect some resemblance to other wines from the same grape but not enough to call a California Pinot Noir "Burgundy." However, because of uniformity of climate, varietal California wines are excellent and are usually equal to if not better than their average French counterparts.

Another rule of thumb in selecting a good wine for table wines (unfortified) is to make sure that it has a cork that must be removed with a corkscrew. This is small point but will help you separate bad domestic wine from the better quality.

Chapter 2

Since there is so much mention of seasonal harvest of wheat (e.g., winter, spring) and place of origin, I have included a chart.

List of Important Wheats and Harvest

Wheat	Harvest Time	Gluten (Percent)	Properties
English	August	7–11	Ordinary varieties produce a flat loaf, but with good flavor.
Australian	January	8–13	Some flours possess soft, putty-like gluten but absorb water fairly well. There are improved varieties as strong as Manitoban. These are not exported, however.
N. American, Minnesota	July	11–15	Produces very strong flour, requires thorough fermentation.
Canadian, Manitoba	August	11–15	Produces very strong flour, requires thorough fermentation.
Pacifics, California, Oregon, etc.	June	8–10	Ricey, soft, good color, Suitable for cakes and biscuits.
S. American Plate	January	10–11	Variable quality, medium strength. Used as a "filler".
Russian	August	10–12	Variable today—not graded. Similar to lower grade Manitoban.
Indian and Persian	April	10–13	Very dry and flinty. Very stable. Dirty.

Chapter 3

More on Cheese.

The two charts following will serve as a summary of the processes described in the text.

CHART 1
Cheese Makers' Options

Fresh Cheeses:
1. slow coagulation—Petits Suisses
2. fast coagulation—Fromages à la Pie

| | | | With Mold | | With a Rind | | |
			External	Internal	Dry	Washed	Covered with Ashes
	natural drainage	slow coagulation	Neuchâtel St. Marcellin				
		fast coagulation	Coulommiers Brie Camembert		Chèvre	Langres Bourguignon	Vendôme Olivet
Ripened Cheeses	drainage accelerated by	cutting	Carré de l'Est	Gorgonzola Bleu d'Auvergne Roquefort *Bleu du Jura*		Pont l'Évêque	
		cutting & heating		Fourme d'Ambert		Tilsiter	
		cutting heating pressing	Tome de Savoie St. Nectaire			St. Paulin Edam Reblochon *Edam*	
		cutting heating pressure milling			Cantal	Cheddar Chester	
		cutting milling cooking pressure			Spalen Parmesan Asiago	Emmenthal Gruyère Comté	

CHART II
Differences in Processing and Their Results

Distinctive Processing	Distinctive Characteristics	Typical Varieties of Cheese
Curd coagulated primarily by acid	Delicate, soft curd	Cottage cheese, cream cheese, Neufchâtel
Curd particles matted together	Close texture; firm body	Cheddar, Cheshire
Curd particles kept separate	More open texture	Colby, Monterey, Edam, Gouda
Presence of small amount of copper from copper cheese kettle or vat	Granular texture; brittle body	Parmesan or Reggiano, Romano or Sardo
Stretched curd	Plastic curd; threadlike or flaky texture	Provolone, Caciocavallo, Mozzarella
Bacteria ripened throughout interior with formation of eyes	Gas holes or eyes throughout cheese	Swiss (large eyes) Gruyère and Asiago (small eyes)
Mold ripened throughout interior	Visible veins of mold	Roquefort, Stilton, Gorgonzola
Surface ripened principally by mold	Typical piquant, spicy flavor Edible crust; soft, creamy interior Typical pungent flavor	Camembert, Brie
Surface ripened principally by bacteria and yeast	Surface growth; soft, smooth, waxy body Typical mild to robust flavor	Bel Paese, Brick, Port-Salut Limburger, Münster
Protein of whey or whey and milk coagulated by acid and high heat	Sweetish flavor of whey	Whey cheese; Gjetost, Mysost, Primost

Source: Bruce H. Axler, *The Cheese Handbook* (New York: Hastings House, 1968), p. 182 (Chart I) and p. 183 (Chart II). Reproduced by permission of the publisher.

The Wood-Werkman Reaction—Something for Nothing!

What is heterotrophic CO_2-fixation? Prior to 1935, the nutritional dichotomy between autotrophs and heterotrophs was rather severe, the former were able to fix CO_2 into cell substance with energy derived from sunlight or mineral oxidation while the latter were unable to do so, getting all their carbon atoms from the partial breakdown of organic compounds which at the same time yielded energy.

H. G. Wood and C. H. Werkman are responsible for softening the demarcation between the two nutritional groups. In 1935, they were studying glycerol fermentation by *Propionibacterium shermani* and were disturbed by a seeming inconsistency in their results. In analyzing the fate of the various elements serving as substrate, they seemed to be getting more carbon taken up than was available. After repeated trials assured them their results were for real, they set about looking for reasons. The source of the extra carbon turned out to be the calcium carbonate ($CaCO_3$) in the medium which served to neutralize the acid produced during the fermentation of glycerol:

$$CaCO_3 + \text{fermentation acid} \longrightarrow CaO + CO_2$$

It was the CO_2 released in the above reaction that was being incorporated (fixed) into the cell. This fact was subsequently confirmed with the use of $CaCO_3$ containing Carbon-14 (radioactive) which was traced to a particular compound in the cell.

CO_2 fixation by heterotrophs was a fact, the difference between autotrophs and heterotrophs in this regard being the source of the energy needed for its fixing (in reality, reduction). Wood and Werkman and others soon demonstrated the universality of this reaction (now called the Wood-Werkman Reaction) by demonstrating it in algae, fungi, protozoa and a variety of animal tissues.

Chapter 4

I thought some of you might be amused by this rather lengthy quasi-official scheme for classifying pickles (R. Binsted et al., *Pickle and Sauce Making,* 2nd ed. London: London *Food Trade Press,* 1962).

Pickle Classification

I. *Dill Pickles*
A. Fermented dill pickles:
    1. Genuine dill pickles.
    2. Genuine Kosher dill pickles.

3. Overnight dill or fresh fermented dill pickles.
4. Overnight or fresh fermented Kosher dill pickles.
5. Polish dill pickles.
B. Unfermented dill pickles made directly from fresh cucumbers:
   1. Fresh or pasteurized dill pickles.
   2. Iceberg or quartered dill pickles.
C. Dill pickles made from salt stock:
   1. Processed imitation or summer dill pickles.
   2. Processed imitation or summer Kosher dill pickles.
   3. Pasteurized processed dill pickles.
   4. Pasteurized processed Kosher dill pickles.
II. *Sour Pickles*
A. Sour Pickles:
   1. Plain sour pickles.
   2. Sliced or hot sour pickles.
B. Mixed sour pickles:
   1. Mixed unspiced.
   2. Spiced or hot mixed pickles
   3. Mixed chutney.
C. Relish, chow chow, etc.
III. *Sweet Pickles*
A. Plain sweet pickles:
   1. Standard sweet pickles.
   2. Midget sweet pickles.
   3. Burgherkins.
   4. Slices, chips or wafers.
   5. Candied chips.
   6. Sweet dill pickles.
   7. Bread and butter or old-fashioned pickles.
   8. Peeled pickles.
B. Mixed sweet pickles:
   1. Plain mixed.
   2. Mustard pickles.
   3. Jamaica pickles.
C. Relish or chopped sweet pickles:
   1. Plain relish.
   2. Spread relish.
   3. India relish.
   4. Piccalilli.
   5. Fruit relish.
   6. Mexican relish.
   7. Vegetable relish.

Did You Ever Get Stood on Your Head by Tequila (distilled Pulque)?

Perhaps the best understood of any metabolic degration process is the fermentation of glucose to ethyl alcohol. The path was so well defined as to become a rut. Gibbs and DeMoss, in 1954, found that *Zymomonas lindneri* produces ethyl alcohol from glucose with the same proclivity as yeast cells. Using radioactive glucose they learned that there was one alcohol molecule, which was derived differently from that traditionally observed in yeasts. This metabolic pathway from glucose to ethyl alcohol has been worked out by Entner and Doudoroff and bears their names. It has facetiously been referred to by Wayne Umbreit of Rutgers as "the case of the upside-down Pyruvate" because the number one carbon of glucose becomes the acid portion of pyruvic acid rather than the number three carbon as in traditional acid fermentation. Perhaps this is the reason that for some of us it seems to take less tequila (distilled from the same fermentation that produces pulque) to stand us on our respective heads.

Chapter 5

The following table summarizes protein yields of the Fungi Imperfecti studied by William D. Gray and associates.

Yields and Protein Percentages of Fungi Imperfecti

Culture No.	Form Genus	Mycelium Dry Wt/ Flask Mgm	Economic Coefficient	Crude* Protein Percent	Extracted Protein Percent	T.P.S.† Mgm
I-14	Phoma	484	1.92	30.8	32.0	155
I-174	Pestalotia	608	1.53	27.0	27.9	170
I-39	Gonytrichum	522	1.79	23.1	27.2	142
I-92	Hormiactella	565	1.65	30.8	25.2	142
I-165	Cylindrocephalum	452	2.07	21.7	24.5	111
I-154	Hormodendrum	540	1.73	16.9	24.2	130
I-164	Tetracoccosporium	460	2.02	19.6	23.4	108
I-138	Rhizoctonia	475	1.98	19.2	22.8	108
I-4	Curvularia	583	1.55	24.9	20.4	119
I-15	Phyllosticta	460	2.02	18.6	18.7	86
I-134	Spicaria	482	1.93	22.1	18.1	87
I-145	Macrophomina	582	1.58	17.8	16.9	99
I-13	Pestalotia	590	1.58	17.8	16.4	97
I-58	Colletotrichum	451	2.07	22.1	16.0	72
I-27	Pestalotia	618	1.51	19.9	15.0	99
I-110	Dothiorella	463	2.01	17.9	15.0	70
I-83	Cladosporium	505	1.84	22.7	12.3	62
59	Rhizoctonia	714	1.31	13.4	11.1	79

*Kjeldahl N x 6.25.
† Total Protein Synthesized.
Source: Developments in Industrial Microbiology, Publication of the Society for Industrial Microbiology (Washington, D.C.: American Institute of Biological Sciences, 1964), 5: 189.

Chapter 6

The following chart offers information on most of the antibiotics sold today.
Some Antibiotics Produced Commercially

Antibiotic	Microbial Source	Antibiotic Spectrum	Chemical Type	Route of Administration	Some Commercial Sources
Amphomycin	Streptomyces canus	Gram-positive bacteria	Polypeptide	Topical	Bristol Laboratories
Amphotericin B	Streptomyces nodosus	Yeast; fungi	Polyene	Oral and Parenteral	E.R. Squibb & Sons
Aterrimin	Bacillus subtilis	Gram-positive bacteria		Animal feed	Bioferm Corp.
Bacitracin	Bacillus subtilis	Gram-positive bacteria	Polypeptide	Topical; animal feed	Bioferm Corp.: Commercial Solvents Corp.; Kayakukoseibusshitsu; S.B. Penick; Chas. Pfizer & Co.
Blasticidin S	Streptomyces griseo-chromogenes	Fungi		Agricultural use (rice diseases)	Kaken Kagaku
Candicidin	Streptomyces griseus	Yeast; fungi	Polyene	Topical	S.B. Penick
Cephaloridine	Chemical derivative of 7-aminocephalosporanic acid	Gram-positive and Gram-negative bacteria		Parenteral	Glaxo Laboratories Ltd.; Eli Lilly Co.
Cephalothin	Chemical derivative of 7-aminocephalosporanic acid	Gram-positive and Gram-negative-bacteria		Parental	Eli Lilly Co.; Glaxo Laboratories, Ltd.

Antibiotic	Microbial Source	Antibiotic Spectrum	Chemical Type	Route of Administration	Some Commercial Sources
Chloramphenicol	Chemical synthesis (formerly prpeared from Streptomyces venezuelae fermentation)	Gram-positive and Gram-negative bacteria; Rickettsia		Oral and parental	Parke, Davis & Co.; Sankyo
Colistin	Bacillus colistinus	Gram-negative bacteria	Polypeptide	Topical and parenteral	Kayakukoseibusshitsu; Banyu Seiyaku; Upjohn Co.
Cycloheximide	Streptomyces griseus	Fungi		Agricultural fungicide	
Cycloserine	Streptomyces orchidaceus	Gram-positive and TB bacteria	Amino acid derived	Parenteral (TB infections)	Commercial Solvents; Shionogi Seiyaku; Sumitomo Kagaku
Dactinomycin (actinomycin D)	Streptomyces antibioticus	Gram-positive bacteria; antitumor	Polypeptide	Wilm's disease (systemic)	Merck, Sharp and Dohme
Erythromycin	Streptomyces erythreus	Gram-positive bacteria	Macrolide	Oral and parenteral	Eli Lilly and Co.; Abbott Laboratories; Roussel
Fusidic acid	Fusidium coccineum	Gram-positive bacteria	Steroid	Oral and parenteral	Leo Pharmaceutical Products
Gentamicin	Micromonospora purpurea	Gram-positive bacteria	Carbohydrate derived	Parenteral and topical	Schering Corporation
Gramicidin	Bacillus brevis	Gram-positive bacteria	Polypeptide	Topical	S.B. Penick
Griseofulvin	Penicillium griseofulvum	Fungi		Oral and topical	Glaxo Laboratories, Ltd.; Imperial Chemical Industries, Ltd.

Antibiotic	Microbial Source	Antibiotic Spectrum	Chemical Type	Route of Administration	Some Commercial Sources
Hygromycin B	*Streptomyces hygroscopicus*	Gram-positive and Gram-negative bacteria; Helminths	Carbohydrate derived	Agricultural purposes (anthelmintic)	Eli Lilly and Co.
Kanamycin	*Streptomyces kanamyceticus*	Gram-positive, Gram-negative and TB bacteria	Carbohydrate derived	Parenteral	Bristol Laboratories; Banyu Seiyaku; Meiji Seika
Leucomycin	*Streptomyces kitasoensis*	Gram-positve bacteria		Oral and parenteral	Toyo Jyozo
Lincomycin	*Streptomyces lincolnensis*	Gram-positive bacteria		Oral and parenteral	Upjohn Co.
Neomycins	*Streptomyces fradiae*	Gram-positive, Gram-negative and TB bacteria	Carbohydrate derived	Topical	Chas. Pfizer & Co.; S.B. Penick; E.R. Squibb & Sons; Takeda Yakuhin; Upjohn Co.
Novobiocin	*Streptomyces niveus*	Gram-positive bacteria		Parenteral and oral	Merck Sharp and Dohme; Upjohn Co.
Nystatin	*Steptomyces noursei*	Fungi and yeast	Polyene	Oral and topical	E.R. Squibb & Sons
Oleandomycin	*Streptomyces anti-bioticus*	Gram-positive bacteria	Macrolide	Oral and parenteral	Chas. Pfizer & Co.
Paromomycin	*Streptomyces rimosus*	Gram-positive, Gram-negative and TB bacteria; protozoa	Carbohydrate derived	Oral	Farmitalia; Parke Davis & Co.

Antibiotic	Microbial Source	Antibiotic Spectrum	Chemical Type	Route of Administration	Some Commercial Sources
Penicillins					
Penicillin G	Penicillium chrysogenum	Gram-positive bacteria	Amino acid derived	Oral (buffered) and parenteral	Abbott Laboratories; Banyu Seiyaku; Glaxo Laboratories, Ltd.; Imperial Chemical Industries; Eli Lilly and Co.; Meiji Seika; Nihon Kayaku; Chas. Pfizer & Co.; E.R. Squibb & Sons; Takeda Yakuhin; Toyo Jyozo; Wyeth
Penicillin V	Penicillium chrysogenum	Gram-positive bacteria	Amino acid derived	Oral	Abbott Laboratories; Eli Lilly and Co.; Meiji Seika; Toyo Jyozo; Wyeth
Penicillin O	Penicillium chrysogenum	Gram-positive bacteria	Amino acid derived	Parenteral	Upjohn Co.
Cloxacillin	Chemical derivative of 6-aminopenicillanic acid	Gram-positive bacteria		Oral and parenteral	Beecham Laboratories; Bristol Laboratories
Dicloxacillin	Chemical derivative of 6-aminopenicillanic acid	Gram-positive bacteria		Oral and parenteral	Beecham Laboratories
Methicillin	Chemical derivative of 6-aminopenicillanic acid	Gram-positive bacteria		Parenteral	Bristol Laboratories; Beecham Laboratories; E.R. Squibb & Sons

Antibiotic	Microbial Source	Antibiotic Spectrum	Chemical Type	Route of Administration	Some Commercial Sources
Nafcillin	Chemical derivative of 6-aminopenicillanic acid	Gram-positive bacteria		Oral and parenteral	Wyeth
Oxacillin	Chemical derivative of 6-aminopenicillanic acid	Gram-positive bacteria		Oral and parenteral	Bristol Laboratories; E.R. Squibb & Sons; Banyu Seiyaku
Phenethicillin	Chemical derivative of 6-aminopenicillanic acid	Gram-positive bacteria		Oral and parenteral	Bristol Laboratories; Meiji Seika; Chas. Pfizer & Co.; E.R. Squibb & Sons; Taito Pfizer
Ampicillin	Chemical derivative of 6-aminopenicillanic acid	Gram-positive and Gram-negative bacteria		Oral and parenteral	Beecham Laboratories; Bristol Laboratories Wyeth
Polymyxin B	Aerobacillus polymyxa	Gram-negative bacteria	Polypeptide	Topical	Burroughs-Wellcome; Chas. Pfizer & Co. Rhone-Poulenc
Pristinamycin	Streptomyces sp.	Gram-positive bacteria	Polypeptide	Oral	
Rifomycin SV	Streptomyces mediterranei	Gram-positive and TB bacteria		Oral and parenteral	Lepetit
Ristocetin	Nocardia lurida	Gram-positive bacteria		Parenteral	Abbott Laboratories
Spiramycin	Streptomyces ambofaciens	Gram-positive and Gram-negative bacteria; Rickettsia	Macrolide	Oral	Kyowa Hakko; Rhone-Poulenc

Antibiotic	Microbial Source	Antibiotic Spectrum	Chemical Type	Route of Administration	Some Commercial Sources
Staphylomycin	Streptomyces virginiae	Gram-positive bacteria	Peptide	Oral	Recherche Industrie Therapie
Stendomycin	Streptomyces endus	Gram-positive and Gram-negative bacteria	Peptide	Agricultural uses	Eli Lilly & Co.
Streptomycin	Streptomyces griseus	Gram-positive, Gram-negative and TB bacteria	Carbohydrate derived	Parenteral; agricultural uses	Eli Lilly & Co.; Merck, Sharp and Dohme; Chas. Pfizer & Co.; E.R. Squibb & Sons; Kyowa Hakko; Meiji Seika
Dihydrostreptomycin	Chemical derivative of streptomycin	Gram-positive, Gram-negative and TB bacteria		Parenteral	Eli Lilly & Co.; Merck, Sharp and Dohme; Chas. Pfizer & Co.; E.R. Squibb & Sons;
Tetracyclines Chlortetracycline	Streptomyces aureofaciens	Gram-positive and Gram-negative bacteria; Rickettsiae	Anthracycline	Oral and parenteral	Lederle Laboratories Nopco
6-Demethyl-7-chlortetracycline	Streptomyces aureofaciens	Gram-positive and Gram-negative bacteria; Rickettsiae	Anthracycline	Oral and parenteral	Lederle Laboratories

Antibiotic	Microbial Source	Antibiotic Spectrum	Chemical Type	Route of Administration	Some Commercial Sources
5-Hydroxy-tetracycline	Streptomyces rimosus	Gram-positive and Gram-negative bacteria; Rickettsiae	Anthracycline	Oral and parenteral	Chas. Pfizer & Co.
Tetracycline	Streptomyces aureofaciens; chemical derivative of chlortetracycline	Gram-positive and Gram-negative bacteria; Rickettsiae	Anthracycline	Oral and parenteral	Bristol Laboratories; Lederle Laboratories; Chas. Pfizer and Co.; Rachel Laboratories
Thiostrepton	Streptomyces azureus	Gram-positive bacteria	Polypeptide	Topical	E.R. Squibb & Sons
Trichomycin	Streptomyces hachijoensis	Fungi and yeast	Polyene	Topical	Fujisawa Yakuhin
Tylosin	Streptomyces fradiae	Gram-positive bacteria	Macrolide	Agricultural uses	Eli Lilly & Co.
Tyrothricin	Bacillus brevis	Gram-positive and Gram-negative bacteria	Polypeptide	Topical	S.B. Penick
Vancomycin	Streptomyces orientalis	Gram-positive and TB bacteria		Parenteral	Eli Lilly and Co.
Variotin	Paecilomyces varioti	Fungi and yeast		Topical	Nihon Kayaku
Viomycin	Streptomyces floridae	Gram-positive, Gram-negative and TB bacteria	Amino acid derived	Parenteral	CIBA; Parke Davis & Co.; Chas. Pfizer & Co.

Epilogue

Caterpillar Species Susceptible to *Bacillus Thuringiensis* and Its Relatives.

Scientific Name	Common Name
Achaea janata	Castor semilooper
Achroia grisella	Lesser wax moth
Acleris variana	Black-headed budworm
Adisura atkinsoni	Lablab podborer
Agrotis ipsilon	Black cutworm
Alsophila pometaria	Fall cankerworm
Amorbia essigana	Tortricid leaf roller
Anisota rubicunda	Green-striped maple-worm
Anisota senatoria	Orange-striped oak-worm
Anomis sabulifera	Jute semilooper
Antheraea pernyi	Chinese oak silkworm
Aphomia gularis	—
Archips argyrospilus	Fruit-tree leaf roller
Archips crataegana	—
Archips rosaceanus	Oblique-banded leaf roller
Arctia caja	Tiger moth
Argyrotaenia mariana	Gray-banded leaf roller
Argyrotaenia velutinana	Red-banded leaf roller
Atteva aurea	Ailanthus webworm
Autographa californica	Alfalfa looper
Autoplusia egena	Bean leaf skeletonizer
Bombyx mori	Silkworm
Bucculatrix thurberiella	Cotton leaf perforator
Caenurgina erechtea	Forage looper
Carausius morosus	Stick insect
Carpocapsa pomonella	Codling moth
Carpophilus dimidiatus	Corn sap beetle
Ceramica picta	Zebra caterpillar
Choristoneura fumiferana	Spruce budworm
Choristoneura murinana	—
Clysia ambiquella	—
Colias lesbia	—
Corcyra cephalonica	Rice moth
Crambus bonifatellus	Sod webworm
Crambus sperryellus	Silver-barred lawn moth
Cremona cotoneaster	Cotoneaster webworm

Scientific Name	Common Name
Dasychira pudibunda	—
Datana integerrima	Walnut caterpillar
Datana ministra	Yellow-necked caterpillar
Dendrolimus sibiricus	Siberian silk-worm moth
Desmia funeralis	Grape leaf roller
Diacrisia virginica	Yellow-bear
Diaphania nitidalis	Picklewom
Ennomos subsignarius	Elm spanworm
Ephestia elutella	—
Ephestia kuhniella	Mediterranean flour-moth
Erannis tiliaria	Linden loope
Estigmene acreae	Saltmarsh caterpillar
Etiella zinckenella	Lima-bean pod borer
Eublemma amabilis	—
Euphydryas chalcedona	—
Euproctis crysorrhoea	Brown-tail moth
Eupterote fabia	Coffee hairy caterpillar
Feltia subteranea	Granulate cutworm
Galerucella luteola	Elm leaf beetle
Galleria mellonella	Greater wax moth
Gelechia gossypiella	—
Gnorimoschema operculella	Potato tuberworm
Harrisina americana	Grape leaf skeletonizer
Harrisina brillians	Western grade leaf skeletonizer
Heliothis peltigera	—
Hellula undalis	Cabbage webworm
Hemerocampa pseudotsuga	Douglas fir tussock moth
Hibernia defoliaria	—
Himera pennaria	—
Holcocera pulverea	—
Homeosoma electellum	Sunflower moth
Hylephia sp.	Fiery skipper
Hypera brunneipennis	Egyptian alfalfa weevil
Hyphantria cunea	Fall webworm
Hyponomeuta malinellus	—
Junonia coenia	Buckeye caterpillar
Lambdina fiscellaria lugubrosa	Western hemlock looper
Lambdina fiscellaria somniara	Western oak looper
Laphygma frugiperda	Fall armyworm
Lobesia botrana	Eudemis grape vine moth
Loxostege commixtalis	Alfalfa webworm

Scientific Name	Common Name
Loxostege sticticalis	Beet webworm
Malacasoma americanum	Eastern tent caterpillar
Malacasoma fragile	Great basin tent caterpillar
Malacasoma neustria	Tent caterpillar
Malacasoma pluviale	Western tent caterpillar
Maruca testulalis	—
Melanolophia imitata	—
Nudaurelia cytherea	Pine emperor moth
Nymphalis antiopa	—
Operophtera brumata	Winter moth
Paleacrita vernata	Spring cankerworm
Papilio cresphontes	Orange dog
Papilio philenor	—
Peridroma caucia	Variegated cutworm
Phalonia hospes	Banded sunflower moth
Phryganidia californica	California oakworm
Pieris brassicae	European cabbage butterfly
Platynota sultana	Omnivorous leaf roller
Platyptilia carduidactyla	Artichoke plume moth
Plodia interfunctella	Indian meal moth
Porthetria dispar	Gypsy moth
Prodenia eridania	Southern armyworm
Prodenia litura	Egyptian cotton leaf worm
Prodenia praefica	Western yellow-striped armyworm
Protoparce quinquemaculata	Tomato hornworm
Proxenus mindara	Rough skin cutworm
Pseudaletia unipuncta	Armyworm
Pyrausta nubilalis	European corn borer
Recurvaria milleri	Lodgepole needle miner
Rhyacionia buoliana	European pine shoot moth
Sabulotes caberata	Omnivorous looper
Schizura concinna	Red-humped caterpillar
Sitophilus granarius	Grain weevil
Sitophilus oryza	Rice weevil
Sitotroga cerealella	Angoumois grain moth
Sparganothis pilleriana	—
Spilonota ocellana	Eye-spotted bud moth
Spodoptera exigua	Beet armyworm
Stilpnotia salicis	Satin moth
Thamnonoma wavaria	—
Thaumetopoea pityocampa	Pine processionary caterpillar

Scientific Name	Common Name
Thaumetopoea processionea	Oak processionary caterpillar
Thymelicus lineola	European skipper
Udea rubigolis	Celery leaf tier
Urbanus proleus	Bean leaf roller
Vanessa cardui	—
Vanessa urticae	—

Selected Commercially Available Examples of Bacillus Thuringiensis

Bacillus Thuringiensis Strain (Serotype)	Product	Country	Supplier
I, III, V	Thuricide	U.S.A.	I.M.C.
III	Dipel	U.S.A.	Abbott
I	Bakthane	U.S.A.	Rohm & Haas
I	Agritrol	U.S.A.	Merck
I	Parasporine	U.S.A.	Grain Proc.
I	Biospor	Germany	Hoechst
?	Sporeine	France	L.I.B.E.C.
I	Bathurin	Czechoslovakia	Spolana n.p., Neratovice
V	Entobacterin	U.S.S.R.	Near Moscow

*Source: H.D. Burges and N.W. Hussey, *Microbial Control of Insects and Mites*, (New York: Academic Press, 1971), p. 744.

Some Registered Uses for B. *Thuringiensis* Products In the U.S.A.*

Pest		Crop
VEGETABLE AND FIELD CROPS		
Alfalfa caterpillar	*Colias eurytheme*	alfalfa
Artichoke plume moth	*Platyptilia carduidactyla*	artichokes
Bollworm	*Heliothis zea*	cotton
Cabbage looper	*Trichoplusia ni*	beans, croccoli, cabbage, cauliflower, celery, collards, cotton, cucumbers, kale, lettuce, melons, potatoes, spinach, tobacco
Diamondback moth	*Plutella maculipennis*	cabbage
European corn borer	*Ostrinia nubilalis*	sweet corn
Imported cabbageworm	*Pieris rapae*	broccoli, cabbage, cauliflower, collards, kale
Tobacco budworm	*Heliothis virescens*	tobacco
Tobacco hornworm	*Manduca sexta*	tobacco
Tomato hornworm	*Manduca quinque-maculata*	tomatoes
FRUIT CROPS		
Fruit-tree leaf roller	*Archips argyrospilus*	oranges
Orange dog	*Papilio cresphontes*	oranges
Grape leaf folder	*Desmia funeralis*	grapes
FORESTS, SHADE TREES, ORNAMENTALS		
California oakworm	*Phryganidia californica*	
Fall webworm	*Hyphantria cunea*	
Fall cankerworm	*Alsophila pometaria*	
Great Basin tent caterpillar	*Malacosoma fragile*	
Gypsy moth	*Lymantria (Porthetria) dispar*	
Linden looper	*Erannis tiliaria*	
Salt Marsh caterpillar	*Estigmene acrea*	
Spring cankerworm	*Paleacrita vernata*	
Winter moth	*Operophtera brumata*	

*From information supplied, in part, through the kindness of International Minerals and Chemical Corp., Libertyville, Illinois and Nutrilite Products Inc., Buena Park, California, U.S.A.
Source: H.D. Burges and N.W. Hussey, *Microbial Control of Insects and Mites* (New York: Academic Press, 1971), p. 73.

Nobel Prize Winners in Fields Related to Microbiology

Year	Winner	Achievement
Physiology and Medicine:		
1901	Emil A. von Behring (1854–1917)	Serum therapy, especially for work on diphtheria antitoxin.
1902	Ronald Ross (1857–1932)	Discovery of the life cycle of the malaria parasite.
1905	Robert Koch (1843–1910)	Work on tuberculosis and the scientific development of bacteriology.
1907	Charles L.A. Laveran (1845–1922)	Study of protozoa-caused disease.
1908	Paul Ehrlich (1854–1915) Elie Metchnikoff (1845–1916)	Their work on immunology including the introduction of quantitative methods.
1913	Charles R. Richet (1850–1935)	Work on anaphylaxis.
1919	Jules J.P.V. Bordet (1870–1961)	Discoveries in the field of immunity.
1928	Charles J.H. Nicolle (1866–1936)	Work on typhus.
1930	Karl Landsteiner (1868–1943)	Discovery of human blood groups.
1939	Gerhard Domagk* (1895–1964)	Discovering prontosil, the first sulfa drug.
1945	Alexander Fleming (1881–1955) Howard W. Florey (1898–1968) Ernst B. Chain (1906–)	The discovery of penicillin and its curative properties in infections including those of the heart, syphilis, and certain types of pneumonia.
1951	Max Theiler (1899–1972)	Developing the yellow fever vaccine.
1952	Selman A. Waksman (1888–)	Work in the discovery of streptomycin and its value in treating tuberculosis.
1954	John F. Enders (1897–) Thomas H. Weller (1915–)	Their successful growth of polio viruses in cultures of

Year	Winner	Achievement
	Frederick C. Robbins (1916–)	tissues and discovery of more effective methods of polio detection.
1958	George W. Beadle (1903–) Edward L. Tatum (1909–) Joshua Lederberg (1925–)	Discovery that genes transmit hereditary characteristics. Experiments establishing that the sexual recombination of bacteria results in exchange of genetic material.
1960	F. Macfarlane Burnet (1899–) Peter B. Medawar (1915–)	Their discovery of acquired immunity, that is, that an animal can be made to accept foreign tissues.
1962	Francis H.C. Crick (1916–) Maurice H.F. Wilkins (1916–) James D. Watson (1928–)	Their determining of the molecular structure of deoxyribonucleic acid (DNA) and its significance for information transfer in living material.
1965	Francois Jacob (1920–) Andre M. Lwoff (1902–) Jacques L. Monod (1910–)	Their discovery of the regulatory processes in body cells that contribute to genetic control of enzymes and virus synthesis.
1966	Francis P. Rous (1879–1970)	Discovery of a cancer virus.
1969	Max Delbruck (1906–) Arthur Day Hershey (1908–) Salvador Edward Luria (1912–)	Discoveries concerning the replication mechanism and the genetic structure of viruses; hailed as having set the solid foundation on which modern molecular biology rests.
1971	Earl Sutherland (1916–1974)	Cyclic AMP in bacteria and higher forms; how hormones work.
1972	Gerald Edelman (1929–) Rodney Porter (1917–)	Structure of immunoglobin.

Year	Winner	Achievement
1975	David Baltimore (1938–) Renato Dulbecco (1914–) Howard Martin Temin (1934–)	Potential viral cause of cancer.
Chemistry:		
1945	Artturi I. Virtanen (1895–)	Work on plant synthesis of nitrogen compounds; development of a new method of making silage.
1946	James B. Sumner (1887–1955) John H. Northrop (1891–) Wendell M. Stanley (1904–1971)	Their preparation of enzymes and virus proteins in pure form.
1961	Melvin Calvin (1911–)	Establishing the chemical reactions that occur during photosynthesis.

*Prize declined under political pressure; received in 1917.

NOTES

Prologue

1. Clifford Dobell, *Anthony Van Leeuwenhoek, and his "Little Animals"* (New York: Dover Publications Inc., 1960), pp. 243–34.
2. The following correspondence took place between Dr. Sedillot and Littré. The former proposed the term *microbe* as an all inclusive one. He consulted Littre, who answered him on February 26, 1878: "Dear colleague and friend, *microbe* and *microbia* are very good words. To designate the animalculae I should give preference to *microbe* . . . Later he wrote, "It is true that microbios and macrobios probably mean in Greek, short lived and long lived . . . the question is not what is most purely Greek, but what is the use made in our language of the Greek roots" (R. Vallery-Radot, *The Life of Pasteur,* p. 267).
3. "When I use a word," Humpty-Dumpty said in a rather scornful tone, "it means just what I chose it to mean—neither more nor less" (Lewis Carroll,*Through the Looking Glass).
4. Once the true nature of viruses was accepted (a definition which included sizes too small for filter retention) the redundancy of "*filtrable* virus" was evident.
5. Many mycologist (biologists who study fungi) divide this group into more than four classes. However for our purposes, the four should suffice.
6. There are only minor differences between chitin and cellulose but sufficient to make them digestible by different enzymes. Chitin is found in certain fungal cell walls as well as in insect

exoskeletons. The bacterial substance is a modified chitin which is responsible for their rigidity.

7. *Bergey's Manual of Determinative Biology,* ed. R.E. Buchanan and M.E. Gibbons (Baltimore: Williams and Wilkins, 1974).

8. N. V. Pirie, "The Meaninglessness of the Terms of Life and Living," *Perspectives in Biochemistry,* ed. J. Needham and D. Green (Cambridge, Mass.: Cambridge University Press, 1937), p.12.

9. This tiny organism which so often preys upon higher life forms is in turn victimized by its own parasites: "So naturalists observe, a flea / Hath smaller fleas that on him prey / And these have smaller fleas to bite 'em / And so proceed ad infinitem." J. Swift, "On Poetry: A Rhapsody."

10. Cultures of specific types or species are available from several sources. The major supplier of bacteria, algae, fungi, and protozoa in this country is the American Type Culture Collection, 12301 Parklawn Drive, Rockville, Maryland 20852. A catalogue of strains is available on request. In addition, there are numerous institutions and laboratories that maintain cultures for research and industry throughout the world (see Appendix).

Chapter 1

1. Recently, a Biblical scholar, H. Hirsch Cohen, has offered some interpretations on this aspect of Noah's behavior in *The Drunkennness of Noah,* (University of Alabama Press, 1974). Noah's nakedness was in preparation for a sexual union and the wine he imbibed an aphrodisiac. Noah drank freely to restore his sexual power. When Ham gazed on his father's uncovered phallus he symbolically stole from Noah his potency and authority as God's agent on earth.

2. Maynard A. Amerine and Ralph E. Runkee, "Microbiology of Winemaking," *Annual Review of Microbiology* 22 (1968): 323–58.

Chapter 2

1. Recent archeological findings (Wendorf, *et al., Science* 169 [1970]: 1161–70) from the Late Paleolithic era along the Nile

suggests that ground grain was used as a source of food as early as 13,000 B.C., 4,000 years sooner than previously thought.

2. Recently Dr. Leo Kline of the USDA has isolated an organism he has named *Lactobacillus San Francisco,* which he claims will make it possible to produce the inimitable sourdough all over the world, which heretofore has not been possible.

3. According to one source, which I haven't been able to verify, this bread was invented by a German baker, one Pumper Nickel. Dolores Casella, *A World of Breads* (New York: D. White & Co., 1966), p. 82.

Chapter 3

1. J.G. Davis, *Basic Technology:* vol. 1, *Cheese* (New York: American Elsevier Publishing Co., Inc., 1965), p. 72. Some of you may notice as you read through this chapter, the glaring omission of so-called fresh cheeses such as cottage cheese and related products, for example, buttermilk. Since by the above definition these do not quality as cheeses they are being treated in Chapter IV.

2. The abomasum or fourth or real stomach in ruminants is the source of rennet. See Chapter VIII.

3. London: Faber & Faber, 1965.

4. To the traveler seeking these caves, a warning: there are three Roqueforts! Please go to Roquefort-sur-Soulzon which is in the Gorges-du-Tarn region of southern France, about 65 miles northwest of Montpellier.

5. I can vouch for this temperature in the caves. It is really cold. Anyone touring this cave facility, where all of the processing takes place, is given a wraparound blanket to wear. The tour lasts the better part of one hour and I personally appreciated the need for cover.

Chapter 4

1. As do *eau de vie* (French) and *akvavit* (Danish), both of which are used euphemistically for the national spirit drink.

2. Memorable indeed! Traditionally in the Highlands, a host would stand by his door and bid each of his guests a memorable good night with a "Wee Deoch an Dorius," a bit of malt whiskey for the road.
3. In fact, it is the first fruit mentioned by name (Genesis 8:11): [On the 47th day Noah sent a dove from the ark.] "And the dove came in at eventide; and lo, in her mouth an olive leaf freshly plucked."
4. The use of vinegar assuredly is ancient. It is mentioned in the book of Ruth (2:14), "And Boaz said unto her, At mealtime come thou hither, and eat of the bread, and dip thy morsel in the vinegar."

Chapter 5

1. Term coined by Professor Carroll Wilson of the Massachusetts Institute of Technology as a euphemism to avoid the use of terms such as *bacterial* or *microbial cell protein.*
2. Cutting fluids are used during a variety of metal-working operations to cool and lubricate tool surfaces. Because of their organic content they are fair game for many microorganisms. These can reduce the efficiency of the cutting fluid, causing rust, foul odors, and disposal problems.
3. William D. Gray, et al., "Fungi Imperfecti as a Potential Source of Edible Protein," *Developments in Industrial Microbiology,* Proceedings of the Twentieth General Meeting of the Society for Industrial Microbiology (Washington, D.C.: American Institute of Biological Sciences, 1964), 5: 384–89.
4. Henry J. Peppler, "Industrial Production of Single-Cell Protein from Carbohydrates," *Single Cell Protein,* ed. Richard I. Mateles and Steven R. Tannenbaum (Cambridge, Mass.: M.I.T. Press, 1968), p. 241.

Chapter 6

1. "While working with *Staphylococcus* variants, a number of culture plates were set aside on the laboratory bench and examined

from time to time. In the examinations these plates were necessarily exposed to the air and they became contaminated with various microorganisms. It was noticed that around a large colony of a contaminating mould, the *Staphylococcus* cólonies became transparent and were obviously undergoing lysis". Alexander Fleming, "On the Antibacterial Action of Cultures of a *Penicillium* with a special reference to their case in the isolation of *B. influenzae*," *British Journal of Experimental Pathology*, 10 (1929), 226–36. The above statement is a perfect example of what Pasteur meant by "Chance favors the prepared mind."

2. *Broad* is a relative term. Its presumed meaning is that the antibiotic is active against many groups of microbes, for example, bacteria, fungi, and viruses. In actuality, it means activity against a larger number of bacteria.

3. Primarily as stated above, antibiotic residues in food increase the chances for selection of resistant strains of microbes, as well as the potential for allergy induction.

4. Just a note to remind some cooks that they cannot make gelatin desserts with fresh pineapple; bromelin digests the gelatin.

5. "Thou shalt not seethe a kid in his mother's milk" (Exodus 23:19).

6. Invasive factors are those substances produced by microbes that are primarily involved in establishing "beachheads" in the host by breaking down initial lines of defense.

Chapter 7

1. "Pollution: an undesirable change in the physical, chemical or biological characteristics of our land, air and water that may or will harmfully affect human life or that of other desirable species, our industrial processes, living conditions, and cultural assets; or that may as well waste or deteriorate our raw material resources." Perhaps, just as appropriate an alternative: "a resource out of place." *Waste Management.* National Academy of Sciences, National Research Council, National Academy of Sciences Printing & Publishing Office, Publication No. 1400, 1966.

2. The new science of Gnotobiology (from Greek *gnoto*, "known").

3. W. W. Eckenfelder and D. J. O'Connor, *Biological Waste Treatment* (Elmsford, N.Y.: Pergamon Press, 1961), p. 8.

4. There has not been any systematic study of the species in compost. Due to the great variety of composting materials, and the different geographies, the best that we can say is related to the succession of types during composting. Cellulose-digesting fungi, which are heat-tolerant, are succeeded by actinomyces (fungi-like bacteria), and finally by the most heat-resistant of all, bacterial spore-formers.

Epilogue

1. An interesting note: In his painting *The Plague of the Philistines,* the 17th century painter Nicholas Poussin depicted dead rats among the victims. This was more than two hundred years before it was discovered that plague was transmitted from rat to man by the flea.

2. Plague might have spared the defenders then, but during the reign of Justinian, 100 years later, it killed half the population of the Byzantine empire.

3. Fracastoro did not readily relate this disease to carnal knowlege but in his subsequent treatise on contagious disease (*Contagium Vivium*), he did indeed trace the condition of his poor shepherd to venereal misbehavior.

4. Gonorrhea is by far the older of the two. Detection and prevention are mentioned in detail in Leviticus 15:1–12.

5. I must confess that I, too, can be taken in by what I read. *Time,* Dec. 8, 1970, p.96, in discussing venereal disease says, "The condom was thought to be invented by a British physician of the 17th century, a Dr. Condom." This statement is somewhat substantiated by the 18th edition of *Dorland's Medical Dictionary.* Dear reader, I spent many hours searching for an authoritative account to verify those two glib statements: Dr. Condom was nowhere to be found, not even in those obscure places where I discovered Dr. Muffet. I give you this bit of miscellaneous information. Condom is a city in Southwest France in the Gers Department and is the center of the Armagnac Brandy district, second only to Cognac in repute.

6. Carl Sandburg, *Abraham Lincoln: The War Years* (New York: Harcourt, Brace Co., 1939), 2: 117–20.
7. *Trial and Error: The Autobiography of Chaim Weizmann* (New York: Harper & Row, 1949).
8. According to another source, Ralph G. Martin, *Jenny: The Life of Jenny Jerome Churchill,* vol. 2 (Englewood Cliffs, N.J.: Prentice-Hall, 1971), Winston Churchill was at that time minister of munitions in the war cabinet.
9. For example, the nursery rhyme "Ring around the rosy" refers to the circular rash around the nipple, found in the plague.
10. Macbeth, Act IV Scene III, ll. 146–59.
    MacDuff: "What's the disease he [the doctor] means?"
    Malcolm: "Tis called the evil:
    A most miraculous work in this good king;
    Which often, since my here remain in England,
    I have seen him do. How he solicits heaven,
    Himself best knows: but strangely visited people,
    All swoln and ulcerous pitiful to the eye,
    The mere despair of Surgery, he craves
    Hanging a golden stamp about their necks,
    Put on with holy prayers: and 'tis spoken
    To the succeeding royalty he leaves
    The healing benediction. With this strange virtue
    He hath a heavenly gift of prophecy
    And Sunday blessings hang about his throne
    That speak him full of grace.
11. In his diary, the Irish poet, Tom Moore, himself a consumptive, relates a conversation he had with Byron whom he visited in Greece in February 1828. "I look pale," said Byron, looking in the mirror, "I should like to die of a consumption." "Why?" asked Moore. "Because the ladies would all say, 'Look at that poor Byron, how interesting he looks in dying!'" René and Jean Dubos, *The White Plague, Tuberculosis, Man and Society* (Boston: Little, Brown & Co., 1952), p.58.
12. Polyhedra are crystallike inclusions enclosing a number of virus particles, which are produced in the cells of susceptible insect tissues. The polyhedra are very resistant to the environ-

ment and dissolve only in the guts of those insects afflicted, releasing the virus particles.

13. I would like to give credit for this title to a published series of lectures with the same title: A. J. Kluyver and C. B. Van Niel, *The Microbe's Contribution to Biology* (Cambridge, Mass.: Harvard University Press, 1956).

SUGGESTED READINGS

Prologue

Curtis, Helena. *The Viruses*. Garden City, New York: Natural History Press, 1966.

Postgate, John. *Microbes and Man*. Baltimore: Penguin Books Ltd., 1969.

Smith, Kenneth M. *Beyond the Microscope*. Rev. ed., Baltimore: Penguin Books Ltd., 1957.

Chapter 1

Amerine, M.A., and Joslyn, M.A. *Table Wines, The Technology of Their Production*. Berkeley and Los Angeles: University of California Press, 1951.

Cohen, H. Hirsch. *The Drunkenness of Noah*. University, Alabama: University of Alabama Press, 1974.

Marrison, W. *Wines and Spirits*. London: Pelican Books, 1957.

Simon, Andre L., ed. *Wines of the World*. New York: McGraw-Hill, 1962.

Simon, Andre L. *The Commonsense of Wine*. New York: The International Wine and Food Publishing Co., 1966.

Waugh, Alec. *In Praise of Wine and Certain Noble Spirits*. New York: William Morrow & Co., Inc., 1971.

Chapter 2

Bennion, Edmund B. *Breadmaking, Its Principles and Practice.* 4th ed. London: Oxford University Press, 1967.

Jay, James M. *Modern Food Microbiology.* New York: Van Nostrand-Reinhold, 1970.

Kleyn, John, and Hough, James. "The Microbiology of Brewing." *Annual Review of Microbiology* 25(1971): 583.

Pomeranz, Yeshajahu, and Shellenberger, J.A. *Bread Science and Technology.* Westport: Avi Publishing Co., 1971.

Porter, John. *All About Beer.* Garden City, New York: Doubleday & Co., Inc. 1975.

Chapter 3

Davis, J.G. *Basic Technology:* vol. 1, *Cheese.* New York: American Elsevier Publishing Co., Inc. 1965.

Foster, Edwin M. *Dairy Microbiology.* Englewood Cliffs, N.J.: Prentice-Hall, Inc., 1957.

Marquis, Vivienne, and Haskell, Patricia. *The Cheese Book: A Definitive Guide of the Cheeses of the World.* New York: Simon & Schuster, 1964.

U.S. Dept. of Agriculture, Dairy Products Laboratory. *Cheese Varieties and Descriptions.* Washington, D.C.: Agriculture Handbook no. 54, 1953.

Chapter 4

Binsted, R., et al. *Pickle and Sauce Making.* 2d ed. London: London Food Trade Press, 1962.

Dwidjoseputro, Dakimah, and Wolf, Frederick T. "Microbiological Studies of Indonesian Fermented Foodstuffs." *Mycopathologia et Mycologia Applicata* 41(1970): 211–12.

Hesseltine, C.W., et al. "Investigations of Tempeh, an Indonesian Food." *Developments in Industrial Microbiology* 4(1963): 275.

Lockhart, Sir Robert Bruce. *Scotch, The Whiskey of Scotland in Fact and Story.* 3d ed. London: Putnam, 1966.

"Microbiological Aspects of Brewing," *Developments in Industrial Microbiology* 4(1963): 153–80.

Pederson, Carl S. *The Microbiology of Food Fermentations*. Westport, Conn.: Avi Publishing Co., Inc., 1971.

Shibasaki, K., et al. "Miso. II Fermentation." *Developments in Industrial Microbiology* 2(1961): 205.

Chapter 5

Bhattacharjee, Jnanendra K. "Microorganisms as Potential Sources of Food." *Advances in Applied Microbiology* 13(1970): 139–59.

"Microorganisms as Potential Food Sources." *Developments in Industrial Microbiology* 7(1966): 203–30.

Chapter 6

Davies, R. "Microbial Extracellular Enzymes, Their Uses and Some Factors Affecting Their Formation." *Biochemistry of Industrial Microorganisms*, Chap. 4: 68–150. New York: Academic Press, 1963.

Duddington, C.L. *Microorganisms as Allies: The Industrial Use of Fungi and Bacteria*. New York: Macmillan Co., 1961.

Dulaney, Eugene L. "Microbial Production of Amino Acids." *Microbial Technology*, Chap. 12: 308–43. New York: Reinhold Publishing Corp., 1967.

Goldberg, Herbert S. "Nonmedical Uses of Antibiotics." *Advances in Applied Microbiology* 6(1964): 91–118.

Goodwin, T.W. "Vitamins." *Biochemistry of Industrial Microorganisms*, Chap. 5: 151–205. New York: Academic Press, 1963.

Haas, G.J. "Some Applications of Enzymes of Microbial Origin to Foods and Beverages." *Food Product Development*, October 1971.

Hughes, Donald H. "Enzymes as Additives in Detergents." *Developments in Industrial Microbiology* 12(1971): 9–58.

Peterson, J.K. "Therapeutic Dentrifrices." *Advances in Applied Microbiology* 13(1970): 343–61.

Sizer, Irwin W. "Enzymes and Their Applications." *Advances in Applied Microbiology* 6(1964): 207–26.

Vining, L.C. and Taber, W.A. "Alkaloids." *Biochemistry of Industrial Microorganisms,* Chap. 10: 341–78. New York: Academic Press, 1963.

Chapter 7

Alexander, M. "Biodegradation: Problems of Molecular Recalcitrance and Microbial Fallibility." *Advances in Applied Microbiology* 7(1965): 35–77.

Alexander, Martin. *Introduction to Soil Microbiology.* New York: John Wiley & Sons, Inc., 1961.

Eckenfelder, W.W., Jr., and O'Connor, D.J. *Biological Waste Treatment.* New York: Pergamon Press, 1961.

Hungate, Robert E. *The Rumen and Its Microbes.* New York: Academic Press, 1966.

Luckey, Thomas D. "Effects of Microbes on Germfree Animals." *Advances in Applied Microbiology* 7(1965): 170–218.

Rae, S.N., and Block, S.S. "Experiments in Small-Scale Composting." *Developments in Industrial Microbiology* 3(1962): 326–40.

Rosebury, Theodor. *Life on Man.* New York: Viking Press, 1969.

Silverman, Melvin P., and Erlich, Henry L. "Microbial Formation and Degradation of Minerals." *Advances in Applied Microbiology* 6(1964): 153–206.

Updegraff, David M. "Microbiological Aspects of Solid Wastes Disposal." *Developments in Industrial Microbiology* 13(1972): 9–56.

Epilogue

Dennie, Charles Clayton. *A History of Syphilis.* Springfield, Ill.: Charles C. Thomas, 1962.

Redford, Myron H., Duncan, Gordon W., and Prager, Denis J., ed. *The Condom: Increasing Utilization in the United States.* San Francisco: San Francisco Press, Inc., 1974.

Rosebury, Theodor. *Microbes and Morals: The Strange Story of Venereal Disease.* New York: Viking Press, 1971.

Zinsser, Hans. *Rats, Lice and History.* Boston: Little, Brown & Co., 1935.

General Reference

Gaden, Elmer L., Jr. *Global Impacts of Applied Microbiology II, Second International Conference on Global Impacts of Applied Microbiology.* New York: Interscience Publishers, 1969.

Starr, Mortimer P., ed. *Global Impacts of Applied Microbiology, Proceedings of a Coordination Conference.* New York: John Wiley & Sons, Inc., 1964.

Tannahill, Reay. *Food in History.* New York: Stein & Day, 1973.

INDEX

Harold W. Rossmoore is professor of biology in the Department of Biology at Wayne State University. He is the author of many articles concerning microbiology, and helps edit several scientific journals. The manuscript was edited by Shirley M. Horowitz. The book was designed by Mary Primeau. The typeface for the text is Optima designed by Hermann Zapf about 1958. The display face is Bauhaus.

The text is printed on International Bookmark paper and the book is bound in Columbia Mills' Fictionette cloth over binders' boards. Manufactured in the United States of America.